Principles of Genomics
and Proteomics

Principles of Genomics and Proteomics

Rakeeb Ahmad Mir

Department of Biotechnology, School of Life Sciences, Central University of Kashmir, Ganderbal, Jammu & Kashmir, India

Sheikh Mansoor Shafi

The Department of Advanced Centre for Human Genetics, Sher-i-Kashmir Institute of Medical Sciences, Soura, Srinagar, Jammu & Kashmir, India

Sajad Majeed Zargar

Proteomics Laboratory, Division of Plant Biotechnology, Sher-e-Kashmir University of Agricultural Sciences & Technology of Kashmir (SKUAST-Kashmir), Srinagar, Jammu & Kashmir, India

ELSEVIER

Elsevier
Radarweg 29, PO Box 211, 1000 AE Amsterdam, Netherlands
The Boulevard, Langford Lane, Kidlington, Oxford OX5 1GB, United Kingdom
50 Hampshire Street, 5th Floor, Cambridge, MA 02139, United States

Notices
Knowledge and best practice in this field are constantly changing. As new research and
experience broaden our understanding, changes in research methods, professional
practices, or medical treatment may become necessary.

Practitioners and researchers must always rely on their own experience and
knowledge in evaluating and using any information, methods, compounds, or
experiments described herein. In using such information or methods they
should be mindful of their own safety and the safety of others, including parties
for whom they have a professional responsibility.

To the fullest extent of the law, neither the Publisher nor the authors,
contributors, or editors, assume any liability for any injury and/or damage to
persons or property as a matter of products liability, negligence or otherwise, or
from any use or operation of any methods, products, instructions, or ideas
contained in the material herein.

ISBN: 978-0-323-99045-5

For information on all Elsevier publications visit our
website at https://www.elsevier.com/books-and-journals

Publisher: Stacy Masucci
Acquisitions Editor: Andre G. Wolff
Editorial Project Manager: Kristi Anderson
Production Project Manager: Kiruthika Govindaraju
Cover Designer: Miles Hitchen

Typeset by TNQ Technologies

Working together
to grow libraries in
developing countries

www.elsevier.com • www.bookaid.org

Contents

About the authors xiii
Foreword xvii
Preface xix
Acknowledgments xxi

**CHAPTER 1   Undestading the OMICS techniques: an introduction
to genomics and proteomics 1**
1 Introduction to genomics 1
  1.1 Structural genomics 2
  1.2 Functional Genomics 3
  1.3 Epigenetics 5
2 Genomes of the eukaryotes—nuclear and organelle genomes 7
  2.1 Chromosomes: a brief account 7
  2.2 The nuclear genome 9
3 Organization of eukaryotic genome 10
  3.1 Structure of genes in eukaryotes 10
4 Organelle genome editing 18
  4.1 The plastome-chloroplast DNA 19
  4.2 Mitome-mitochondrial DNA 19
5 Introduction to proteomics 20
  5.1 The protein structure and overview to the structural
      hierarchies 21
6 Proteome profiling 25
7 Conclusion 26
  References 26

CHAPTER 2   Genome mapping 29
1 Introduction 29
  1.1 Linkage maps 29
  1.2 Physical maps 30
2 Markers 30
  2.1 Simple-sequence repeats 31
  2.2 Inter simple sequence repeats 32
  2.3 Randomly amplified polymorphic DNA 33
  2.4 Restriction fragment length polymorphism 35
  2.5 Amplified fragment length polymorphism 36
  2.6 Single nucleotide polymorphism 37
3 Physical mapping 40
  3.1 Restriction mapping 40

3.2 Fluorescence in situ hybridization....................................41
3.3 Sequence-tagged site mapping 44
References...44

CHAPTER 3   Analysis of genomes—I.....................................47
1 Introduction 47
   1.1 Genome sequencing 47
2 Sangers sequencing/chain-termination sequencing method..........48
   2.1 The basic principle of chain-termination DNA sequencing... 48
3 Whole-genome shotgun sequencing..50
4 Hierarchical shotgun sequencing..50
5 Next-generation sequencing methods.....................................51
   5.1 Major NGS techniques include......................................51
   5.2 The sequencing methods included in next-generation
       sequencing..53
6 Genome sequencing projects...55
   6.1 The human genome project 55
   6.2 Some facts of HGP 57
   6.3 Arabidopsis genome..59
   6.4 The introduction to microbial genome sequencing.............59
   6.5 Concept of genome annotation......................................60
7 Conclusion...62
References...63

CHAPTER 4   Analysis of genomes—II....................................65
1 Introduction 65
2 DNA footprinting/deoxyribonuclease I (DNase I) protection
   mapping 66
   2.1 Workflow of DNA footprinting.......................................67
   2.2 Applications of DNA footprinting...................................69
3 Gel retardation assay or electrophoretic mobility shift assay
   or band shift assay...69
   3.1 Basic working principle 70
   3.2 Variants of EMSA..71
   3.3 Applications of EMSA..73
4 Chromatin immunoprecipitation and its variants......................73
   4.1 Workflow of ChIP in detail...74
   4.2 The derivatives of ChIP technique 77
   4.3 The ChIP databases...80
   4.4 Applications of ChIP..81
5 Conclusion...83
References...83

CHAPTER 5 Nutrigenomics: Insights into the influence of nutrients on functional dynamics of genomes..........89

1 Introduction 89
1.1 The beginning of nutrigenomics.................................. 92
1.2 Nutrigenomics research tools 92
1.3 Markers for nutrigenomics: the assessment of single nucleotide polymorphism.. 93
1.4 Human nutrition and diseases.................................... 94
1.5 Celiac disease.. 94
1.6 Irritable bowel syndrome.. 95
1.7 Gastroesophageal reflux disease................................ 95
1.8 Food allergies.. 96
1.9 Oral disease 96
1.10 Lifestyle-associated diseases 96
2 Nutrigenomics studies of selected crop plants—a repository of functional foods..97
2.1 Buckwheat... 99
2.2 Quinoa (*Chenopodium quinoa* wild)......................... 99
2.3 Amaranth (*Amaranthus* spp.) 100
2.4 Plant-based foods to address the "lifestyle" diseases 100
2.5 Role in regulating hypertension................................ 101
2.6 Food having antidiabetic properties........................... 101
2.7 Nutrigenomics, lifestyle diseases, and food quality go hand in hand.. 102
2.8 Nutrigenomics for improving crop............................ 102
2.9 Technical intervention: biofortification of crops through nutrigenomic techniques 103
3 Conclusion.. 104
References.. 105

CHAPTER 6 Analysis of proteomes—I 111

1 Polyacrylamide gel electrophoresis and SDS-PAGE 111
1.1 Principle of PAGE 111
1.2 Basic requirements and workflow of PAGE.................... 113
1.3 Buffers used in PAGE... 114
1.4 The gel 114
1.5 Preparation of protein samples 115
1.6 Running the gel... 116
1.7 Staining of gel and visualization of proteins.................. 116
1.8 Gel interpretation.. 116

2 Isoelectric focusing gels 117

2.1 Preparation of IEF gels 117

3 2D gel electrophoresis-combining SDS-PAGE and IEF for unison gel-based identification of proteins on the basis of size and charge 118

3.1 Basic principle of 2D-GE 118

3.2 Procedure for setup the 2DGE 118

4 MALDI-TOF mass spectrometry 120

4.1 The principle and setup of MALDI-ToF 120

4.2 Basic instrument and working setup 121

4.3 The most commonly used matrix compounds in MALDI ToF 122

4.4 Laser types used in MALDI-ToF 122

4.5 Liner TOF analyzer and reflectron TOF analyzer 122

4.6 The reflectron for higher resolution 124

4.7 Applications of MALDI ToF 124

4.8 Quality control 125

5 Isotope-coded affinity tags 125

5.1 Basic principle and workflow 125

5.2 Applications of isotope-coded affinity tag 127

6 Isotope-coded protein labeling 127

6.1 Applications 128

7 Isobaric tags for relative and absolute quantification 128

7.1 Basic principle and workflow 129

7.2 Some advantages of iTRAQ 130

7.3 Some disadvantages of iTRAQ 130

7.4 Applications 130

8 The tandem mass tag 131

9 Stable isotopic labeling of amino acids in cell culture 131

9.1 Principle of SILAC 132

9.2 The technical workflow of SILAC 132

9.3 SILAC-based applications 133

9.4 Studies of secretomes 133

9.5 Characterization of proteins on the basis of posttranslational modifications 134

9.6 Studying cancer cell proteomes 134

10 Discussion 135

References 135

CHAPTER 7 Analysis of proteomes—II **139**
   1 Introduction 139
   2 Ultraviolet and visible light spectroscopy 139
      2.1 Basic principle.. 140
      2.2 Major factors that affect the absorption by UV/Vis
           spectrum.. 141
      2.3 Basic workflow.. 143
      2.4 The monochromator 143
      2.5 Basics for analysis of samples 144
      2.6 Some strengths/limitations 145
      2.7 Applications of UV–visible spectrophotometer............ 145
      2.8 Quantification and purification of DNA and RNA
           analysis 145
      2.9 Analysis of pharmaceuticals............................. 146
      2.10 Analysis of beverages 146
      2.11 Checking the number or concentration of bacterial
           cells in cultures... 146
      2.12 Applications.. 146
   3 Fluorescence spectroscopy................................... 146
      3.1 Basic principle 146
      3.2 Basic workflow for fluorescence detection and its
           measurement.. 147
      3.3 Major applications of fluorescence spectroscopy 148
      3.4 Utility of fluorescence spectroscopy..................... 149
   4 Nuclear magnetic resonance spectroscopy 149
      4.1 The basic principle... 150
      4.2 About magnetic field 151
      4.3 NMR spectroscopy of proton 151
      4.4 Case study—1: H_2O molecule........................... 152
      4.5 The concept of chemical shift............................. 152
      4.6 The signal strength... 153
      4.7 The π-electron functions in NMR......................... 153
      4.8 The significance of NMR 154
      4.9 Applications of NMR....................................... 154
      4.10 NMR for elucidating the metabolite structures 154
   5 X-ray diffraction.. 154
      5.1 The crystallization of proteins............................. 155
      5.2 The optical setup in X-ray crystallography................ 156

5.3 The mechanism of diffraction analysis........................... 156
5.4 Crystal structure determination................................... 157
5.5 The rotating crystal method..................................... 157
5.6 Retrieving and processing of data............................... 158
5.7 Applications of X-ray diffraction method....................... 158
6 Circular dichroism..159
6.1 Basic workflow of CD... 160
6.2 Applications... 161
7 Conclusion... 161
References... 161

CHAPTER 8   Analysis of proteomes—III **165**
1 Protein structure prediction.................................... 165
2 Protein secondary and tertiary structure predictions................. 165
3 Homology or template-based modeling.............................. 166
3.1 Template identification and initial alignment 166
3.2 Generation of models 167
3.3 Modeling of loops... 168
3.4 Model assessment 168
4 Protein threading... 169
4.1 Software for protein threading................................... 169
5 Ab initio protein structure predictions............................. 170
6 Conclusions 173
References... 173

CHAPTER 9   Analysis of proteomes—IV **177**
1 Introduction.. 177
2 Yeast 2-hybrid system 177
3 Principle... 179
4 Advantages and limitations 180
5 Limitations of Y2H assays.. 180
6 Phage display 180
7 Application 181
8 Protein chips 182
9 Applications of protein microarrays............................... 184
9.1 Analytical microarrays...184
9.2 Functional protein microarrays..................................185
10 Conclusion and future prospective................................ 187
References... 188

CHAPTER 10   Beyond genomics and proteomics......................191

    1 Introduction 191

    2 Transcriptomics 192

       2.1 Technological intervention.. 192

    3 Metabolomics...196

       3.1 Technical intervention 197

       3.2 Workflow for metabolome analysis 199

       3.3 Applications of metabolomics...................................... 200

    4 Interactomics..204

       4.1 Methodology of interactomics...................................... 205

       4.2 Applications of interactomics....................................... 208

    5 Lipidomics...209

       5.1 Technical intervention 210

       5.2 Applications of lipidomics.. 211

    6 Metagenomics 213

       6.1 Methodology employed for metagenomics 214

       6.2 Applications of metagenomics 215

    7 Ionomics...218

       7.1 Technical intervention for studying ionomics.................. 219

       7.2 Applications of ionomics.. 221

    8 Connectomics..221

       8.1 Technical inputs for connectomics................................ 222

       8.2 Applications of connectomics 224

    9 Conclusion...225

    References...226

    Further reading..242

Index...245

About the authors

Dr. Rakeeb Ahmad Mir
Assistant Professor
Department of Biotechnology
School of Life Sciences
Central University of Kashmir, Ganderbal,
J&K, India

Dr. Rakeeb Ahmad is currently an Assistant Professor at the Department of Biotechnology, School of Life Sciences, Central University of Kashmir, Ganderbal, India. He is having 06 years of research and teaching experience in different subjects of biotechnology. He has worked as an Assistant Professor at Baba Ghulam Shah Badshah University, Rajouri (BGSBU) in India. His research interests include molecular taxonomy, nematode genomics, and basic research in proteomics. He is currently involved in publishing research and review articles in international journals and has served as peer reviewer for several reputed journals. In addition, he is deeply involved in teaching courses such as, cell biology, signal transduction and cancer biology, molecular biology, genomics and proteomics, and nematode genomics at university level. He has mentored several students for carrying major research-based projects at master's level. Dr. Rakeeb has published various research and review papers, book chapters on molecular taxonomy of nematodes, proteomic techniques, and stress tolerant crop plants.

Dr. Sheikh Mansoor Shafi
Postdoctoral Researcher
mansoorshafi21@gmail.com;
mansoorbiochem@gmail.com
https://orcid.org/0000-0003-1902-4611

Dr. Sheikh Mansoor Shafi is a postdoctoral researcher, presently working at the SK-Institute of Medical Sciences Soura Srinagar. Previously, he was working as project associate-I at the CSIR—Indian Institute of Integrative Medicine Jammu (IIIM). He has obtained his BSc from SP College of Science (University of Kashmir), MSc from HNBGU Uttarakhand, and a Ph.D. degree from Sher-e-Kashmir University of Agricultural Sciences and Technology of Jammu. Dr. Mansoor has made important contributions to his field and has been working hard to shape himself as an impressive scientist of the highest intellectual caliber. He has an impressive lab experience in biochemistry, molecular biology, and genetics to his credit. He has collaboratively worked and published more than 30 research and review papers in highly reputed journals like Frontiers, Plos one, Scientific reports, and JoF MDPI. He has published more than 10 book chapters and three books, and he has filed one patent application, which was published on 01/04/2022. Dr. Mansoor has delivered many guest lectures at different institutes and seminars/workshops. He has submitted more than 200 sequences to NCBI GenBank. He has attended several national and international conferences and workshops like European Molecular Biology Organisation (EMBO) and also received hands-on training on Advance Techniques in Modern Biology and Advanced Bioinformatics, Genomics, and NGS data analytics. He is the recipient of many awards like the Best Researcher Award, Best Thesis Award, and Best Poster Award. Dr. Mansoor is a review editor in Frontiers in Ethnopharmacology and is also the reviewer of several reputed journals like 3 biotech, Cell Reports, Frontiers, Plosone, and MDPI.

Sajad Majeed Zargar
Assistant Professor
Sher-e-Kashmir University of
Agricultural Sciences and Technology
of Kashmir (SKUAST-Kashmir), India

Sajad Majeed Zargar, Ph.D., is currently
an assistant professor at Sher-e-Kashmir
University of Agricultural Sciences &

Technology of Kashmir (SKUAST-Kashmir) in India and visiting Professor at University of Padova, Italy. He was previously a visiting professor at the Nara Institute of Science & Technology, Japan. He has worked as an assistant professor at SKUAST-Jammu, Baba Ghulam Shah Badshah University, Rajouri (BGSB) in India. He has also worked as a scientist at Advanta India Limited, Hyderabad, India and TERI (The Energy & Resources Institute), New Delhi, India. Dr. Zargar is in receipt of a CREST overseas fellowship from DBT, India; Goho grant from Govt. of Japan, and Erasmus Fellowship from European Commission. He has received several awards for his work and research. He is also the member and representative of INPPO (International Plant Proteomics Organization). He has chaired a session in 3rd INPPO, World Congress held at University of Padova, Italy in 2018. He is a member of various International and national scientific societies. His editorial activities and scientific memberships include publishing research and review articles in international journals and as a reviewer. He has been affiliated with several internationally reputed journals and is also a reviewer of reputed journals Journal of Advanced Research, Frontiers in Plant Science, 3 Biotech, Scientia Horticulture, Methods in Ecology and Evolution, Australian Journal of Crop Science, and many others. He has also edited several books that are published by international reputed publishers. Dr. Zargar has been invited to give many lectures at professional meetings and workshops and has received grants for research projects under his supervision. He is presently coordinator/principal investigator of rice genomics and buckwheat genetics projects funded by DBT, New Delhi, India. He has supervised many M.Sc. and Ph.D. students. He has been the mentor of three Post Doc Fellows funded by SERB, New Delhi and DST, New Delhi. He is the in-charge Scientist of Genomics Lab. and Proteomics Labs. at Division of Plant Biotechnology, SKAUST-Kashmir, India.

Foreword

Prof. Nazir A. Ganai
Vice-Chancellor

Sher-e-Kashmir
University of AgricultureSciences and Technology of Kashmir
www.skuastkashmir.ac.in

OMICS is a fast-advancing field that broadly refers to high-throughput approaches such as genomics, transcriptomics, proteomics, or metabolomics. Through these approaches, researchers seek to study all the genes and other functional moieties in the cell of an organism simultaneously. With the recent progress in genome sequencing of various organisms, the information that guides biological function and development lies at the heart of these changes, which is why genomics has become a central and consistent discipline of research. A great landmark in genomics was the Human Genome Project (HGP), which provided the first glimpses into the entire human genome in 2000. On 26 June 2000, President Bill Clinton in a meeting stated that humanity had discovered nothing less than the language of God. The completion of HGP led to several genome projects that included personalized diploid human genome sequencing. Due to these advances, we now talk about the human digitome, a digital setup that is built to provide data for personalized and predictive medicine for human health and the prevention of diseases.

Although genomics has been of tremendous importance, the functionality of genes expressed under specific conditions in a particular tissue remained a challenge, and thus led to the field of proteomics. The progress made in proteomics in the last 2 decades is unbelievable with the technological advances in proteomics involving a shift from gel-based to gel-free proteomics techniques like ICAT, iTRAQ, etc., along with the recent technological advances in mass spectrometry that have made it possible to trace the expression of low abundance proteins that has been a challenge. Given the progress made in various omics tools and their applications, I believe this book entitled, *"Principles of Genomics and Proteomics"* will be of great importance. The book provides a concise and clear account of the technical aspects of genomic and proteomic analysis as well as their application to biological systems. Furthermore, an in-depth explanation of the fundamental principles in understanding genomes and their structural and functional aspects is covered. As both genomics and proteomics fields are fast advancing and complementing each other for elucidating molecular mechanisms, etc., the book provides the basic concepts involved in understanding its significance and applications. The chapters covered are designed in a well-defined chronology that will help readers in understanding the technical as well as application parts. This book will be of tremendous importance to the undergraduates and postgraduate students of biotechnology, biochemistry, medicine, agriculture, and chemistry. Moreover, it will be of great help to the researchers using OMICS as a tool in their research.

I applaud the authors for their hard work in compiling this useful book with relevant topics.

(Nazir A. Ganai)

Place: Shalimar
Dated: 10-10-2022

Preface

Since the last few decades, advances made in molecular biology have helped in an in-depth understanding of various biological processes that regulate genetic networking. OMICS-based technologies that include genomics, transcriptomics, proteomics, metabolomics, and ionomics have revolutionized the field of molecular biology. For translational research, omics have been of great importance. With the availability of genome sequences of various organisms, it has been possible to understand the functionality of various genes that were unknown. Complete genome sequences of model organisms followed by comparative genomics and functional genomics have enabled us to understand the genetic regulations among different organisms.

This book *Principles of Genomics and Proteomics* covers various topics related to genomics, nutritional genomics, proteomics, etc. The book provides a brief and clear account on the technical aspects of genomic and proteomic analysis and their applications. Moreover, it covers other topics like structural and functional genomics, etc. All the chapters are well designed for easy understanding by the readers.

This book will be of great importance to the undergraduates and postgraduate students of biotechnology, biochemistry, medicine, agriculture, and chemistry. Moreover, it will be of great help to the researchers using OMICS as a tool in their research.

Rakeeb Ahmad Mir
Sheikh Mansoor Shafi
Sajad Majeed Zargar

Acknowledgments

With the blessing from Almighty Allah, we could accomplish his book on time. It would have been difficult without the support and encouragement of many people associated with us professionally or unprofessionally. We would like to thank our mentors who have always been a motivation for us and our students and whose research using OMICS techniques has been a great inspiration to start this project. I, Dr. Rakeeb Ahmad Mir, am highly indebted to Prof. Mohammed Afzal Zargar, Registrar, of the Central University of Kashmir, for his valuable suggestions, continuous support, and encouragement. I would also like to thank Dr. Sajad Majeed Zargar (Co-author) for initiating this project by developing a roadmap for designing this proposal and for his valuable input in the compilation of this book. I, Dr. Sheikh Mansoor Shafi, would like to pay sincere appreciation to Dr. Arshad A. Pandith, Dr. Javid Iqbal Mir, and Dr. Parvaiz Ahmad, for their consistent guidance and support throughout my career. I count myself lucky for having had your mentorship. I am forever grateful to my parents for their love, support, guidance, and kindness. I, Dr. Sajad Majeed Zargar, acknowledge DBT, New Delhi, India for financial support to undertake research on the genomics of colored rice of Jammu and Kashmir, India, and buckwheat genetics. I, Sajad Majeed Zargar, also acknowledge the European Commission for sanctioning the Erasmus + Project that involves staff and student mobility between the University of Padova, Italy, and SKUAST-Kashmir, India, to undertake research on crop OMICS. The authors are also grateful to the Hon'ble Vice Chancellor, SKUAST-Kashmir (Prof. Nazir Ahmad Ganai), for the support, encouragement, and guidance.

This book is dedicated to all those people who change the lives of others by living as an example and empowering others.

Undestading the OMICS techniques: an introduction to genomics and proteomics

1. Introduction to genomics

Living organisms on earth are specified by the genomes of a huge number of diversified organisms. We now know very critically that all organisms possess a genome that stores biological information needed for normal growth and metabolism. The genomes of organisms are made of DNA (deoxyribonucleic acid), and RNA as a viral genome in viruses is an exception. Both DNA and RNA are constructed of nucleotides held by phosphodiester bonds. The DNA molecule is constructed of two polynucleotide chains held by hydrogen bonds between nucleotides in a helical fashion described by Watson and Crick in 1953. The last few decades of the genomic era led to the deciphering of huge information pertaining to the construction and composition of genomes by the application of genomic technologies. By employing automated sequencing technologies and computer algorithms, the genomes were used to sequence and assemble a large amount of nucleotide sequences of humans and several model organisms. All the DNA sequences/nucleotides-based data are organized and stored by National Center for Biotechnology Information, European Molecular Biology Laboratory, National Institute of Health, and DNA Data Bank of Japan. All these biological databases are continuously updated, and they exchange data on a global scale through the internet. Out of this collaborative approach, a large number of genomic sequences are determined related to a large number of organisms such

Principles of Genomics and Proteomics. https://doi.org/10.1016/B978-0-323-99045-5.00001-X

as bacteria, viruses, archaea, plants, and animals. Several techniques have been dedicated to investigate the genomes of living organisms, which include sequencing technologies such as automated and next-generation sequencing techniques. Researchers examine and mine the treasure trove of DNA/protein data to deeply investigate gene functions, identifying new genes, evolutionary relationships, and expression analysis by employing bioformatics tools (Lodish, 2016).

We are now at the helm of a deep understanding of genomics, that is, studying the whole genomes and the genetics of organisms. The study of genomics includes the application of recombinant DNA technology, different sequencing technologies, and the use of bioinformatics tools to assemble, analyze, and interpret the structure and gene functions. In addition, genomics grossly helps to study the interaction between alleles and loci in genomes. Moreover, genomic studies also help to study heterosis, epistasis, and pleiotropy. Out of these investigations, scientists further divided genomics into structural genomics, functional genomics, epigenomics, and metagenomics. All these fields are used to study genomes at diverse levels to investigate the genes coding for proteins, description of gene functions and their interaction. In addition, OMICs approaches also help to study a complete set of epigenetic modifications, and investigate the genetic material directly derived from diverse habitats by metagenomics.

1.1 Structural genomics

This field of genomics helps to elucidate the functions and interactions of genes and proteins by the application of genome-wide genome studies. The studies rely on various processes such as hotspots of DNA, transcription, translation, and protein functions based on protein−DNA and protein−protein interactions (Fig. 1.1). Based on later studies, dynamic networks and models are constructed to unravel the precise nature of expression patterns in quiescent, differentiated, and dividing cells. Several technological advances aided in improved gene annotations, genome-wide studies to understand molecular, and interactions in cells and their organelles. The experimentations employed in functional genomics involve measuring total changes in DNA or RNA, interaction of total proteins with DNA or RNA that collectively influence the phenotype. The major branches studied under functional genomics include genotyping and epigenetic

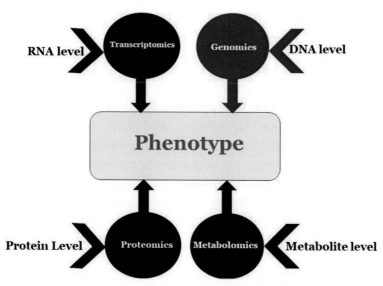

FIGURE 1.1

Functional genomics based on genomics, transcriptomics, proteomics, and metabolomics to investigate particular phenotype of organism under study.

profiling, transcription profiling, DNA/RNA−protein interaction profiling, and metagenomic profiling.

1.2 Functional Genomics

The study of overall DNA structure and its organization of an individual genome helps to manipulate DNA segments, specifically genes of a particular organism. Nowadays, the structural levels of a large number of genomes have been characterized by employing comparative genomic studies. Further studies help to elucidate the genes and markers of every chromosome, gene mapping, physical mapping of chromosomes, and finally by sequencing of genomes. The important objectives of structural genomics are to deduce the genomic structure, protein functions, deduction of protein folding, and identification of critical sites in proteins for the discovery of drugs. The process of structural genomics involves genes and markers that are assigned to an individual chromosome. It is followed by uncovering the structural details of chromosomes and then sequencing the target genes. Several techniques are employed to determine the structural features of genomes such as de novo methods, which involve cloning, and expression of all the Open Reading Frames (ORFs) of the whole genome. Later process was followed by purification and crystallization of expressed proteins and analysis by nuclear magnetic resonance and X-ray crystallography. Ab initio modeling is another technique to predict the 3D structure of proteins. The predicted proteins are then compared by protein homology to decipher the structure of unknown proteins by sequence-based modeling technique. Finally, the threading method is used to find out the similarities in structural modeling and folding pattern of unknown proteins. It is evident that structural genomics is grossly aimed to exploit the recent flood of sequence information and expression patterns to elucidate the precise structure of proteins, which further aids in structure-aided drug designing. To be precise structural genomics is a collection of bioinformatics tools to elucidate the high-throughput 3D structural determination and critical analysis of biomolecules (Fig. 1.2).

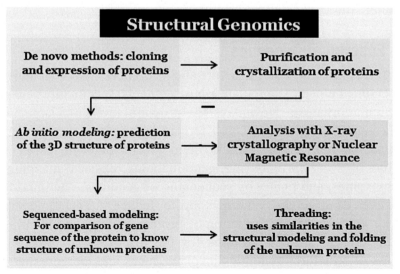

FIGURE 1.2

Hierarchy of process involved in structural genomics.

1.3 **Epigenetics**

It is an important field of genomics, which includes studying the heritable fluctuations in the expression of genes without changing the nucleotide sequences of DNA molecules. So it is the second level of genetic control in addition to the natural genetic codes of DNA, due to which genes are turned off or on as per the requirement of organisms. The epigenetic changes largely rely on the chemical modifications of DNA or histone proteins in a cell. These changes act as potential markers to make genes active or inactive. These modifications are further briefly explained in the following section.

1.3.1 At DNA level: DNA methylation

In this type of modification, a methyl group is added to the DNA by the enzyme DNA methyltransferases. The modification occurs throughout the DNA at promoter regions, specifically on the 5-position (C5) of cytosine nucleotides found adjacent to the guanine nucleotides to form 5-methyl cytosine. Multiple CpG are found in clusters and are usually named as CpG islands. In contrast, unmethylated CpG promoters have low-affinity histone octamers. The array of CpG nucleotides is found in promoters of genes encoding proteins that are not required in large quantities.

1.3.2 At protein level
1.3.2.1 Histone modifications

Chromatin is a nucleoprotein structure, which organizes the DNA in the nucleus of eukaryotic cells. Chromatin is associated with both replication, inactivation and activation of transcription, and DNA repair, for the proper regulation of several fundamental processes. Chromatin consists of nucleosomes as fundamental units, and each of the nucleosome's core is consisting of an octamer of histones (H2A, H2B, H3, H4 and variants), with 146/147 bp DNA wrapped around it (Luger et al., 1997). These histones consist of histone tails that are subjected to several modifications such as methylation, acetylation, phosphorylation, ubiquitination, SUMOylation, and ribosylation. There are multiple types of histone modifications, which are catalyzed by a number of enzyme families; the most well-characterized modifications include acetylation and methylation of histones H3 and H4. The modifications directly alter the DNA—

protein interactions to change how chromatin is structured, which will alter the ability for a gene to be transcribed and expressed.

- Acetylation:

The acetylation of histones is mediated by histone acetyltransferases, to add an acetyl group to lysine amino acids in the tails of histones, such that the positive charge of amino acid is masked, hence lessening the affinity of DNA with histones and resulting in loosening of chromatin for gene activation (Strahl & Allis, 2000). The acetylation process is reversed by enzyme histone deacetylases, which remove the acetyl groups on lysine and hence condensation of chromatin and inactivation of genes.

- Methylation

The process of methylation usually occurs on lysine or arginine amino acids in histones and is catalyzed by histone methyltransferases. This enzyme mediates mono-, di-, or tri-methylation events on target histones. This class of epigenetic mark does not change the charge of the target amino acid and results in both activation and inactivation of genes (Hayakawa & Nakayama, 2011).

- Histone phosphorylation:

This class of posttranslational modification occurs at threonine, serine, and tyrosine amino acids of histone tails, critical for the remodeling of chromatin. The phosphorylation of H2A histones demarcates the critical domains of chromatin in the state of DNA damage of cellular response and serves as the best example. Phosphorylation and ubiquitination are reported to play a critical role in the regulation of chromatin.

- Ubiquitination:

The process of ubiquitination is accomplished by tagging ubiquitin to target protein destined for degradation by 26S proteasomal degradation in the cytosol or through clastosomes in the nucleus. Moreover, ubiquitination is also involved in diverse processes such as immune response, antigen processing, protein regulation, control of cell cycle, and most specifically chromatin regulation (Eytan et al., 1989). The major roles played by ubiquitination include epigenetic silencing via the polycomb repressive complex, bookmark ubiquitination, transcription elongation, and DNA repair. Whereas, most histones are targets of ubiquitination for chromatin remodeling (Swatek & Komander, 2016). The histone undergoes monoubiquitination rather than poly-ubiquitination to modify chromatin. The process of ubiquitination is of paramount importance to the functioning of cells and its association with dreadful diseases such as cancers.

- SUMOylation:

This class of histone modification is accomplished by small ubiquitin-like modifier (SUMO) proteins, which are similar to ubiquitin. A variety of SUMO protein isoforms that are conjugated to substrates almost in a similar fashion as found in the process of ubiquitination have been reported (Jürgen, 2004). Critical to its mechanism is it helps in the SUMOylation of histones at lysine residue, later tagging results in its ability to be modified by multiple sets of enzymes. For example, SUMOylation is found to be frequently associated with the repression of genes through its conjugation with transcription factors such as c-Jun, Sp3, Elk1, and c-Myb. The histone H4 is the primary candidate protein targeted by SUMOylation at 5 lysine residues in its amino-terminal tail (Dafna et al., 2003). It is evident that histone modifications are critical to the expression of genes and play a key role in development, metabolism, physiology, and cellular dynamics.

2. Genomes of the eukaryotes—nuclear and organelle genomes

2.1 Chromosomes: a brief account

Eukaryotic cells organize DNA molecules in a highly organized order within the nucleus. As the length of DNA is a hundred thousand times as that the diameter of a cell, it is important to package DNA into constricted architecture. Moreover, it is essential to prevent knotting or tangling of DNA molecules during the division of cells, for precise segregation to the daughter cells. To make these critical tasks possible, several abundant proteins are at the forefront. Usually, DNA is associated with histones and nonhistone proteins organized in the nucleus of interphase cells or nondividing cells. Each individual linear chromosome consists of only a single DNA molecule. It is reported that the longest DNA in humans has approximately 10 cm in length and comprises 2.8×10^8 bp of nucleotides. As mentioned earlier, nucleosomes are critical to the packaging of DNA in the nucleus. Apart from the presence of histones such as H1 (linker histone), H2A, H2B, H3, H4, other histone variants like H1.0–H1.5, H1oo, H1t, and H1× are also found in humans chromosomes for special functional and structural purpose. For higher order folding of chromosomes, the nucleosomes are linked by linker histones to form chromatosomes. A single linker histone is attached to each nucleosome, to form the chromatosomes to form 30 nm fiber in association with nonhistones proteins such as cohesion and condensins. The chromosomes are further condensed up to a diameter of 700–750 nm during the metaphase of the cell cycle.

Within the nucleus, the chromosomes are arranged into chromatins in compacted form and it is this form, which is always available for replication, recombination, repair, and transcription, depending on the type of gene segment and signal received by a particular type of cell. We have discussed the recent understanding of chromatin through the intervention of biotechniques employed in the field of cell biology. We now understand that many genomic functions are regulated by the 3D architecture of chromatin in the nucleus (Cavalli & Heard, 2019; Dekker & Mirny, 2016; Yu & Ren, 2017). Several robust techniques such as genome architecture mapping (Beagrie et al., 2017) and Hi-C (Lieberman-Aiden et al., 2009) have revealed chromatin exists as domains and compartments. Most recently, the multiplexed error-robust fluorescence in situ hybridization is used to trace and demonstrate the imaging of >1000 loci on genome and synthesis of new transcripts from different regions of the chromosome; in addition, scientists have successfully remodeled the architecture of nuclear structures (Su et al., 2020). Markedly, the chromatin is subdivided into topologically associated domains (TADs), the regions with higher interactions. TADs are regions in genomes where coregulated genes exist and harbor regulatory epigenetic elements (Dekker & Mirny, 2016; Krijger & de Laat, 2016; Yu & Ren, 2017). Moreover, it is reported that human genomes are subdivided into ~2000 TADs, and interaction among these TADs is strong and very rarely weak in-between fewer TADs. TADs are further divided into sub-TADs upon analysis by higher (4 kb) resolution using 3C carbon copy (5C) methods in genomes of mouse embryonic stem cells. These sub-TADs are found to interact with proteins such as CTCFs, Med12 (mediator complex component), and SMC1 (structural maintenance of chromosomes 1) (Rao et al., 2014). The adjacent TADs are separated by highly transcribed DNA regions called as CTCF protein binding sites. CTCF protein, a CCCTC-binding factor (CTCF) is a zinc finger protein transcription factor, and its interactions result due to longer-range interactions between topological domains of chromatin and the CTCF loops have been attributed to establish the interactions between topological domains. The importance of CTCF proteins has been demonstrated by the deletion of the CTCF site by CRISPR/Cas9 editing within Hox clusters resulted in the disruption of topological boundaries in mouse ES cells, the deletion resulted in the expression of HOX genes. Further, at a larger scale, it is reported that chromatin subdivided into A and B compartments consists of active and inactive chromatin regions and the critical physiological significance lies in its dynamic changes during the development and in-between different cell types (Lieberman-Aiden et al., 2009; Yu & Ren, 2017). The DNA sequences found in the A compartment harbor active gene regions, that is, they are actively involved in transcription, even though a few regions are not transcribed. In contrast, compartment B harbors inactive DNA gene segments where transcription is suppressed, even though reports suggest very few genes are transcribed in these regions (Lieberman-Aiden et al., 2009).

2.2 The nuclear genome

The nuclear genome is split into a set of different linear DNA molecules that vary from species to species existing in the form of chromosomes. The number of chromosomes varies from one to several hundred per cell. For example, in *Myrmecia pilosula* (ant), only one chromosome is found, whereas 04 chromosomes are found in Indian muntjac deer. The functional segment of DNA, that is, genes are distributed unevenly throughout the genome. As far as the human genome is concerned, the length of 24 DNA molecules varies from the shortest, which is 1.6 cm, to the longest DNA having an 8.5 cm length. So it is evident that a highly organized packing system is employed by cells to reduce the length of DNA length. The total length of DNA in a cell measures 2 m in length, to be adjusted in <10 μm diameter nuclei. Cells have evolved specialized proteins, which associate with nuclear DNA.

3. Organization of eukaryotic genome

The genome is the sum of total genetic material of an organism, which stores biological information. The genomes of eukaryotic organisms contain DNA as genetic material, and viruses have both RNA and DNA as genetic material. The genetic material of eukaryotes consists of nuclear and organelle genomes (mitochondria and chloroplast). The genome consists of both coding and noncoding sequences. The coding sequences of DNA are known as genes and are functional regions that code for protein-coding RNAs, other RNAs, and regulatory sequences (Fig. 1.3). The following subsections provide a lucid account on the different aspects of eukaryotic genomes.

3.1 Structure of genes in eukaryotes

3.1.1 The genes and their related sequences

The regions of DNA that code for proteins consist of exons, transcribed into mRNA by the process of transcription and posttranscriptional modifications. The complete set of exons found in a genome is termed as exome that can be translated into proteins during protein synthesis. Moreover, a complete set of DNA segments that code for a protein and comprise initiation and termination codons is called as ORF. It is evident in eukaryotes that each mRNA is translated into a single protein and expressed from separate transcription units. It is reported that the transcription units are classified into two classes, *viz.*, simple and complex. As far as simple transcription units are concerned, they lead to the synthesis of only one protein product, such as synthesis of β-globin gene. However, these types of transcription units are rare. In contrast, complex transcription units are most

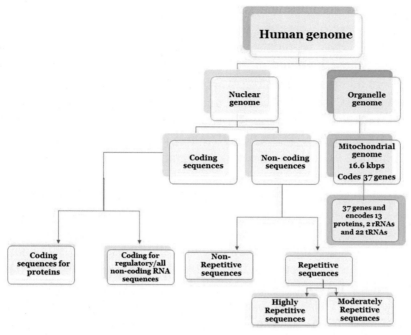

FIGURE 1.3

Overview for organization of human genome.

commonly found in eukaryotic genomes. In this class of transcription units, primary transcripts are processed in more than one way, resulting in the formation of multiple mRNAs. Multiple mRNAs are usually formed by alternate splicing mechanisms and it is this phenomenon, which results in expanding the number of proteins in eukaryotes. Mutation in complex transcription units will create serious problems in an organism because each transcription unit controls the synthesis of several proteins.

Two classes of genes are found based on function and structure. The first class includes solitary genes, which have only one copy in the genome. For example, the chicken lysozyme gene, which cleaves polysaccharides in the cell wall of bacteria. Duplicated genes are the second class of genes, which consists of multiple copies in the genome and are close but nonidentical. These closely related homologous proteins comprise a protein family such as the immunoglobulin superfamily, olfactory receptors, and kinases. During evolution, a segment of a chromosome is duplicated in plants and animals; as a result today we observe the occurrence of gene families and pseudogenes. The DNA segments encoding rRNAs and noncoding RNAs, especially those involved in RNA splicing exist in tandemly repeated arrays. In addition, there are several noncoding RNAs, such as small nuclear RNAs (snRNAs), small nucleolar RNAs (snoRNAs), microRNAs (miRNAs), telomeric RNA, 7SL RNA, long noncoding RNAs such as XIST RNA, TUDOR RNA, and HOTAIR RNA are found in the genomes, which are involved in regulation of genome expression and other functions.

Based on gene density, chromosomal DNA consists of gene-rich regions where a few hundred bps separate transcription units. In contrast, gene-poor regions also called gene deserts, in which intergenic regions are separated by a few million bps. After sequencing the human genome, it was found that 2.9% corresponds to exons, out of which 1.2% encode proteins. It is thought that about 55% of human genomic DNA is transcribed into different types of RNAs and 95% of these RNAs are intronic in nature. Subsequently, it is evident that about 97% of human DNA does not code for genes and is entirely noncoding in nature.

3.1.2 Regulatory elements found in genome
3.1.2.1 Promoters
They are short sequences found immediately to the upstream region of the start site of transcription and are functionally divided into core promoter and proximal promoter. Moreover, promotors are the binding sites for RNA polymerase in combination with initiation factors and combinely the transcription is initiated from the start site. DNA stretch to be transcribed largely depends on the choice of promoter and it is the most frequently employed regulatory site for gene expression. The core promoters initiate the basal transcription of genes. It is the minimal set of DNA sequences critical for the accurate initiation of transcription by RNA polymerase II and associated machinery. It is typically 40−60 nucleotides long and may span in both the upstream and downstream of the start site of transcription. The core promoters include a TATA box/elements recognized by the TATA-binding protein subunit of TFIID (Transcription Factor II D) found located at −25 upstream and GC box, initiator element (Inr),

DPE or Downstream Promoter Element, DCE or downstream core element, and MTE or motif ten element. Proximal promoters are another class found located immediately upstream of the core promoter, usually starts from −50 to 200 bps.

3.1.2.2 CpG islands

Transcription of genes with promoters containing a TATA box or initiator element begins at a well-defined initiation site. However, the transcription of most protein-coding genes in mammals (∼70%) occurs at a lower rate than at TATA box-containing and initiator-containing promoters and begins at any of several alternative start sites within regions of about 100−1000 bp that have an unusually high frequency of CG (cytosine and guanine) sequences.

3.1.2.3 Other important regulatory elements

Also called, as upstream activating sequences, DNA elements are found in complex eukaryotic organisms rather than simple organisms like yeast. They serve to enhance the basal level of transcription. They contain multiple short control elements usually located from 200 bp to tens of kilobases upstream or downstream from a specific promoter. Other elements are silencers, which are DNA elements that trigger the reduction or repression of genes. They are further divided into classical silencers and position-independent silence elements. They are differentially located in the DNA around the promoter sites. The boundary elements also called as insulators are DNA segments, which span 500 bp−3000 bp in DNA and are important for blocking the transcription of genes. In simple terms, they inhibit the spreading of the effect of silencers and enhancers. In addition, responsive elements modulate the transcription of genes against external stimuli. They are located 1 kb upstream of the transcription start site. They respond to second messengers such as cAMP.

3.1.2.4 Other noncoding sequences

Several noncoding functional RNAs are coded by genomes that include miRNA, siRNA, TUDOR RNA, HOT AIR RNA, Xist RNA, Sca RNA, and 7SL RNA. These RNA molecules play a critical role in a diverse range of processes.

3.1.2.5 The introns

They are noncoding sequences, which are found in both gene and non-gene segments of the genome. The introns are replaced by spliceosome complex in mRNAs, whereas introns in tRNA are replaced by proteins. Four distinct groups of introns are reported in eukaryotes based on splicing patterns.

3.1.2.6 UTRs

The noncoding sequences that are found on both sides, that is, 5′ side, it is called the 5′ untranslated region (UTR) (leader sequence) and 3′ side, it is called the 3′ UTR (or trailer sequence) of functional mRNA. The length of 5′ UTR may be >100 nucleotides, whereas 3′ UTR may have a length of several kbs.

3.1.2.7 The pseudogenes

Dysfunctional of a gene, which has lost the ability to be transcribed or if transcribed will lead to the synthesis of dysfunctional protein. These classes of segments are the result of multiple mutations, and their existence usually oes not create any problems for the host organism. Two types of pseudogenes are found in eukaryotic genomes, *viz.*, unprocessed pseudogenes, which lack introns and consist of only exons, and processed pseudogene, which consists of all normal parts of normal genes.

3.1.2.8 Intergenic/extragenic DNA

Noncoding DNA sequences located between genes are known as intergenic regions. Possible functions attributed to intergenic sequences are to control genes, but most of these regions have no known functions and are sometimes known as junk DNA. They encompass about 75% of the genome in humans.

3.1.2.9 Nonrepetitive and repetitive DNA sequences

These two types of genomic sequences are identified by reassociation kinetics. Nonrepetitive or single-copy sequences have only one copy in the haploid genome. Whereas, repetitive sequences consist of multiple copies in the haploid genome. These sequences have a rapid rate of renaturation kinetics. Moreover, the proportion of both types of sequences varies among species. Eukaryotes consist of both repetitive and nonrepetitive DNA sequences, whereas prokaryotes consist of majorly nonrepetitive DNA sequences. Repetitive DNA sequences comprise 80% of the genome in eukaryotes. Therefore, it is evident that repetitive DNA sequences increase the size of the genome. The repetitive DNA is further classified as highly repetitive and moderately repetitive explained below:

3.1.2.10 Highly repetitive DNA sequences

The component of genomic DNA that renatures most rapidly and consists of short sequences <100 bps repeated several times in tandem repeats and is few a times called as simple sequence repeats. Upon shear degradation of human genome, the fragment can be separated on the basis of buoyant density, as followed by CsCl gradient centrifugation. The latter technique results in the separation of buoyant densities. Due to the different composition of these simple sequence repeats and in turn difference in buoyant densities, distinctive bands called satellite bands are formed against the main band during CsCl density gradient centrifugation. It is concluded that the term simple sequence repeats is synonymous with satellite DNA, and they are usually found in eukaryotes. Several classes of satellite DNA sequences, such as alphoid and beta satellite DNA found in centromere, minisatellites and microsatellites found in telomeres and dispersed throughout the genome, respectively. The simple sequence DNAs are mostly located in the heterochromatic regions and as such, they are not transcribed. Another class of satellite DNA that includes microsatellite and minisatellite DNA occurs in tandem repeats, even though they are not found in the satellite band. Minisatellites exist in clusters of up to 20 kb and are 20–50 times repeat at one region of DNA. For example, the telomeric sequences in human 5′TTAGGG3′ repeats almost hundred times in telomeres of human chromosomes. Minisatellites are known for higher mutation and higher rate of diversity even among populations of a species. Moreover, they are also considered "hotspots" of recombination.

In contrast, microsatellites also called as simple sequence repeats or short tandem repeats are clusters of up to 150 bps and existing repeated unit may be 13 bps or less and are 1, 2, 3, or 4 bp units repeated from 10 to 20 times at one region of DNA. They comprise up to 2% of genomes and are thought to be the result of replication slippage.

3.1.2.11 Moderately repetitive DNA sequences

Also known as intermediate repeats or interspersed repeats are DNA sequences, which are found 10 to 1000 times in the genome and are very short in length. These sequences are also dispersed throughout the genome and a large proportion of these sequences are represented by transposons. They comprise about 25%−50% of genomic DNA, and in human's up to 45% in humans. Critical to their functions is that these sequences are able to move from one region to another region in the DNA. Due to later property, these elements are called as transposons or mobile DNA elements. These sequences are found abundantly in eukaryotes and very less frequently in prokaryotes. The process of copying the segment of DNA from one place to other is called as transposition. As these elements have no critical functions, Francis Crick has named them as selfish DNA. Therefore, they are second to the mechanisms of meiotic recombination to bring the chromosomal recombination during the process of evolution. It must be pointed out that defective transposons, incapable of transposing in genomic DNA, are now known as junk DNA.

Barbara McClinktock in Mazie (Corn) discovered the first mobile DNA element in the 1940s. She characterized the movement of genetic elements in and out of genes, which resulted in changing the phenotype of the corn kernels. Since then transposons have been explored at great deal and now we have good information on different classes of transposons in eukaryotes. Transposons are found to fall into two categories based on mechanisms of transposition: type I transposons and type II retrotransposons. As far as type I transposons, they operate by copy-paste mechanism and transpose via RNA intermediate and then through cDNA intermediate, and they are again converted to dsDNA by reverse transcriptase. They operate in analogous to retroviral infection. Whereas, type II transposons operate through copy-paste mechanism. They evolve due to mutations and are a major source of diversity and change. Several transposons are found to be inserted within the gene segment and result in the inactivation of gene expression. Moreover, they may block the expression of genes downstream of their insertion due to the presence of transcription or translation termination signals. In eukaryotes, a number of transposons have been reported to meditate diverse classes of attributes.

Barbara McClintok in the 1950s reported the activator (Ac) elements and dissociation (Ds) elements, a class of transposons, which are found associated with chromosome breakage. She also postulated that the mechanism of transposition by the Ds element is dependent on the Ac element and both of these transposons changed their position in the maize genome. Retrotransposons are another class of transposons, which are found too frequently in the eukaryotic genome and they belong to the type I class of transposons. They operate and transpose by involving RNA intermediate by using reverse transcriptase enzyme. Retrotransposons are further divided into LTR (Long terminal repeats)- and non-LTR-dependent retrotransposons, grossly dependent based on the presence or absence of LTR direct repeats.

LTR transposons: They are commonly found in Drosophila, for example, copia elements and yeast, such as Ty elements. They constitute about 8% of the human genome and are marked by the presence of 250−600 bp long LTRs flanking the coding regions of proteins, very prototypical to retroviruses. They encode all the proteins similar to retroviruses except the envelope proteins. As they have a lot of similarities with retroviruses, they are also called as retrovirus-like elements. Especially, they encode reverse transcriptase and integrase critical for retrotransposition. The major steps of the general mechanism involve the formation of RNA copy by the intervention of reverse transcriptase enzyme followed by the formation of DNA from the RNA transcript. Later, the dsDNA linear molecule is integrated into the genome by an integrase enzyme.

3.1.2.11.1 Non-LTR transposons. This class of transposons lacks LTRs, but rather have A/T rich stretch at one of its ends. They are nonviral transposons and are further divided into long-interspersed elements (LINEs) and short-interspersed elements (SINEs). SINEs are about 6 kb long, nonautonomous transposons that require the enzymatic machinery of LINEs for transposition. These nonviral retrotransposons are transcribed by RNA polymerase-III. *Alu* element is the best example of SINEs, and this element contains a restriction site for *Alu* restriction enzyme. These elements constitute about 10% of the human genome. In addition, a similar type of transposons, B1 is identified in mouse. Both Alu and B1 are thought to be evolved from the noncoding 7SL scRNA, involved in growing polypeptides in the RER.

3.1.2.11.2 Long interspersed elements. They are typical retrotransposons, which can move autonomously. Sequencing technologies have identified about 900,000 sites for LINEs in human genome and comprise about 21% of the genome. Several classes of LINEs have been identified, such as L1, L2, and L3. Only L1 among these LINEs are functional transposons currently identified in human genomes. Basic structural features include the presence of short direct repeats, two ORFs coding for proteins—ORF-1 is about 1 kb in length and codes for RNA binding protein, whereas ORF-2, about 4 kb long, encodes a protein having bot reverse transcriptase and endonuclease activity. L1 elements were found to induce mutation into the gene resulting in myotonic dystrophy and hemophilia upon its transposition. The mechanism of transposition, in this case, differs due to the absence of LTRs. RNA intermediate in this mechanism is accomplished by RNA polymerase-II, directed by the promoter sequences at the left site where LINE is integrated. Moreover, the RNA intermediate of LINEs is polyadenylated in similar fashion as normal cellular mRNAs. Once polyadenylated, the RNA intermediate is exported to the cytosol for translation of proteins from ORF1 and ORF2, followed by insertion of dsDNA at new sites in the genomic DNA.

4. Organelle genome editing

Organelle genome editing consists of mitome (mitochondrial DNA) and plastome (chloroplast DNA) are found isolated from nuclear DNA. The gene found in organelles genomes is separately expressed and regulated to support the respective organelles. Both these DNA molecules are circular and may code all or some tRNAs and rRNAs, and even a large number of proteins are exported from the cytosol. In certain organisms such as Paramecium, Chlamydomonas, and several yeasts, linear DNA molecules also exist. The inheritance carried by both the DNA molecules is sometimes called as cytoplasmic inheritance. Animals typically show mitochondrial inheritance patterns, while in plants both mitochondrial and chloroplast inheritance are observed, but at a smaller scale.

4.1 The plastome-chloroplast DNA

The chloroplast DNA is up to 120−160 kb in length, depending on the species. Both linear and circular cpDNAs are observed in plants, even though now it is evident that higher plants harbor linear DNA molecules, head to tail of linear concatamers. Multiple copies ranging from 20 to 40 copies of cpDNA molecules are found in a single chloroplast, even though the number may exceed 90/chloroplast. For example, 80−90 copies of chloroplast DNA/chloroplast and up to 1000 copies in Chlamydomonas. The total number of genes encoded by cpDNA ranges from 87 to 183 and encodes up to 50 proteins. The important genes encoded by cpDNA include rRNAs such as 23S, 16S, 5S, and 4.5S and 30 tRNA molecules. Moreover, cpDNA encodes up to 30 proteins involved in photosynthesis and ATP synthase complex. Even though, nuclear DNA encodes 95% of proteins required for chloroplast functioning.

4.2 Mitome-mitochondrial DNA

The total size of mitochondrial DNA considerably varies from animals to plants. In the case of animals, the mtDNA is approximately 16.6 kbs, and from experimental studies, it is reported that 99.99% of mtDNA is maternally inherited. In plants, mtDNA is exclusively inherited from the female parent, that is, maternally. The number of mtDNA may reach up to several hundred per cell. The gene content may vary among species, ranging from 03 in *Plasmodium falciparum* to 93 in *Reclinomonas Americana*. In humans, each mitochondrion contains 10 mtDNA molecules and with an estimate of up to 8000 mtDNA per cell. The number of proteins coded by mtDNA is up to 13 in mammals and up to 08 proteins in yeast. The mtDNA genome consists of about 37 genes and encodes 13 proteins, 2 rRNAs, and 22 tRNAs. The mitochondrial genes are expressed in a polycistronic manner as occurs in prokaryotes. All the proteins coded by mtDNA are synthesized by translation machinery of mitochondrion, and these proteins are important for oxidative phosphorylation. The genome is not enveloped and does not form a chromatin-like structure. The mutation rates in the mitochondrial genome are far higher than the nuclear genome which is why we find heterogonous diversity of mtDNA in a population. Several clinical disorders are associated with mitochondrial mutations such as liber hereditary optic neuropathy (Man et al., 2003). Moreover, it is reported that mitochondria generate ROS, which in turn may lead to mutation in mitochondrial DNA to induce cancers, aging, and other disorders.

5. Introduction to proteomics

The complete information regarding the dynamics of cells, tissues, organs, and organ systems. Since most of 20th century, researchers focused on the analysis of individual proteins such as analyzing the enzymatic activity of protein enzymes, its requirement of cofactors, substrates, structure, and mechanistic insights. We now know that it is pertinent that individual proteins such as enzymes involved in metabolic pathways or signaling cascades are interdependent on other proteins for structural and functional dynamics. However, now we believe that studying individual proteins at a broader level does not provide the global picture of total proteins and their differential characteristics. Development in newer technologies has led us to have deep insights into the protein components of cells and led to the evolution of the field of proteomics to study proteins in unison. Proteomic techniques involve studying total proteins, that is, proteomes of cells/tissues/organs/organisms to study different aspects of proteins such as sequence analysis, expression analysis, and structural and functional studies (Fig. 1.4) (Twyman, 2013). It is quite prevailing that proteomes play a critical role in lining the genomes and the biochemical ability of cells to survive senescence and diseases. Determination of activity profiles of proteins, that is, protein profiles, and understanding their biochemical characteristics is of dire importance. The information obtained through proteomics not only encompasses the identification of total proteins but also clarifies the functional diversities and their localization in cellular compartments. Technical approaches to investigate the proteome is termed as expression proteomics or protein profiling. The experimental solutions to study the proteomes are profoundly solved

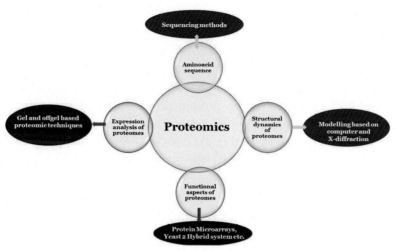

FIGURE 1.4

Diverse fields studied under proteomics to investigate proteomes.

by involving a diverse range of proteomic techniques such as mass spectrometry. To resolve, display, and characterize the mixtures of a large number of proteins, protein biologists for long-used 2D gel electrophoresis. The spots developed through this technique are utilized for the excision of specific proteins and then processed for MS-based analysis. Even though alternative techniques such as high-throughput LC-MS/MS have also been employed to characterize the proteomes. To conclude the proteomic techniques in conduction with molecular genetic methods can precisely and accurately characterize the complex network of proteins in eukaryotes. Such a systematic investigation of protein profiles till date provides clear insights into the architecture of proteins in cells and their changing working modules correlated with the physiological perspectives of cells and organisms.

5.1 The protein structure and overview to the structural hierarchies

Proteins are polymers of amino acids with differential sizes, shapes, and a wide range of functions in the cells. Proteins are synthesized on cytosol of the membrane of RER (rough endoplasmic reticulum) according to the coded sequences carried by processed mRNA, exported from the nucleus. At both sites, ribosomes are at the forefront to translate the information of mRNA into a polypeptide chain, a long polymer of at least 22 standard/proteinogenic amino acids. Once the synthesis of polypeptide begins, chaperones such as HSP70s prevent its misfolding, and finally, the protein is folded by chaperonins into complete functional proteins via a proper hierarchy such as the formation of secondary, tertiary, and in the case of multimeric proteins, quaternary structures are formed. As far as the prevention of misfolding and proper folding in the lumen of RER is concerned, N-glycosylation and involvement of BiP, Calnexin, and Calreticulin are at the vanguard. The principal to the functions of proteins starts after tertiary structure. Moreover, the composition of amino acids and their side chains decide the structural and functional aspects of proteins. Examples of the protein functions include enzymes that catalyze the chemical reactions, signaling proteins and receptors to initiate gene expression of inhibition, regulatory proteins to control the sensors and switches to diverse metabolic and other processes, membrane transporters to carry ions and molecules in and outside the cells, and finally molecular motors, that is, to regulate the movement of vesicles during the process of exocytosis and endocytosis and muscular movements.

There are four levels observed for the formation of proteins viz., primary, secondary, tertiary, and quaternary. The primary structure is the arrangement of amino acids in a linear fashion held by peptide bonds. The most critical element to the primary structure and its ability to form the secondary, tertiary, and quaternary structure depends upon the sequence of amino acids. Whereas, secondary structure refers to the rising structures such as α helical (left and right handed) and β pleated sheets (parallel and antiparallel). Moreover, tertiary structure refers to the 3D folded structure of the polypeptide chain.

5.1.1 The prolog to primary structure of proteins

It is the linear arrangement of proteins synthesized from the coded regions of mRNA by ribosome machinery. At least 22 standard amino acids are required to build a large number of classes of proteins. The backbone of amino acids is held by peptide bonds, a type of covalent bond with the partial double-bind character. Peptide bond is formed in-between the amino group of one amino acid and carboxyl group of another amino acid, with the release of one water molecule and is catalyzed by rRNA within ribosomes. The backbone of proteins is formed by the repeated N, α carbon (Cα), carbonyl C, and oxygen atoms of each amino acid residue. The backbone of proteins exhibits directionality from N to C direction, resulting in one end of proteins with amino-terminal and the other with a carboxyl terminal. Two amino acids are held by one peptide bond and is called as dipeptide, similarly three amino acids are held by two peptide bonds, and so forth. Term oligopeptide is used when a few amino acids are joined by peptide bonds and more than 15–20 amino acids are joined by peptide bonds, the polypeptide term is used for protein. It must be kept in mind that the average molecular weight of amino acid residues is 110–113. This value can be utilized to calculate the molecular weight of proteins.

5.1.2 Central elements of proteins—the secondary structure

The stable spatial arrangement of segments of the polypeptide chain is held together by hydrogen bonds. Multiple secondary structures may be found in a single polypeptide chain grossly dependent on the amino acid constituents/sequence. Major types of secondary structures, *viz.*, alpha (α) helix, the beta (β) sheet, and the short U-shaped beta (β) turn are observed in proteins. Majority of proteins contain about 60% of α-helix and β sheets in their native structure and hence are major supporting secondary structures of proteins.

5.1.2.1 The α Helix secondary structure

The α Helix secondary structure forms a spiral, rigid, rod-like structure formed upon twisting of the primary structure. As far as screw sense is concerned, α Helix exists as both left handed (anticlockwise) and right handed (clockwise). The right-handed α helical secondary structure is energetically more favored than β sheets. Regular α helical structure consists of 3.6 amino acid residues per turn of the helix and the pitch, that is, the distance between two points in the helical axis is 0.54 nm. Each of the residues is related to another residue by 0.15 nm. Each turn of the α helical structure consists of 13 atoms from H to O of the bond and is referred to as 3.6_{13} helix. On an average, the length of α helix is 10–15 amino acids and has a conformation of $\Psi = -47$ and

$\phi = -57$ as per the Ramachandran map. Hydrogen bonds stabilize the α helical structures formed between electronegative carbonyl oxygen of fourth amino acid and nitrogen atoms of peptide linkage. The amino acids showing higher propensities to form α helical structures include lysine, arginine, glutamine, glutamate, alanine, and methionine. Whereas, proline tends to disrupt the α helical structures. The left- and right-handed α Helix secondary structures are measured by employing circular dichroism spectroscopy.

5.1.2.2 The β Helix secondary structure

It consists of laterally packed β strands, and each of β strands has up to 5—8 residues in an extended conformation rather than a coiled structure. The hydrogen bonds are formed between β strands within the single polypeptide chains or in between two polypeptide chains. Each amino acid is about 3.5 Å distant from the adjacent one. They exist as both parallel and antiparallel structures, whereas in the former, the β strands are arranged in the same direction, and in the latter, they are arranged in opposite directions. These structures exist in hydrophobic cores in some proteins, whereas they are embedded in the membranes in a few cases for example porins. They form nearly two-dimensional β pleated structures.

5.1.2.3 Smaller secondary structures—the β turns

They are usually located on the protein surfaces and in the reverse direction leading to the formation of sharp bends. They are stabilized by hydrogen bonds formed between end amino acid residues. Amino acids, proline and glycine, are commonly found in β turns. The built-in bend of proline and lack of side chain in glycine best suit for the existence of β turns at bends of proteins. There are at least three classes of turns, α turn, β turn, and γ turn, based on the separation between the two end residues participating in hydrogen bonds.

5.1.3 The tertiary structure: a startup of functional state

Tertiary structure refers to the 3D structural arrangement of all its amino acid residues in globular proteins. The type of 3D structure formed depends on amino acid sequels of the primary structure. These structures are stabilized by hydrophobic interactions, electrostatic interactions, hydrogen bonds, van der Waals force of interactions, and covalent bonds in the form of disulfide bonds in the case of noncyto-solic proteins formed between cysteine or methionine residues. In addition, threonine and serine are also found on the bends of globular proteins. The core of the tertiary structure comprises nonpolar amino acids, and the surface is composed of hydrophilic amino acids. Myoglobin is a typical example, which led to a clear understanding of tertiary structure. This protein consists of 153 amino acid residues and a single heme group. The interior of this protein consists of nonpolar amino acid residues and the outside consists of polar as well as nonpolar amino acids.

5.1.4 The quaternary structure: a multimeric functional state of proteins

A large number of proteins consists of two or more polypeptide chains or subunits, which may be identical or nonidentical polypeptide chains. They are further divided into two structures fibrous and globular proteins. In fibrous proteins, the polypeptide chains are arranged in parallel in strands or sheets. They have a great role to play in the physiology of animals such as keratins help in the protective and structural aspects of the body. Whereas, in globular proteins, the polypeptide chain is folded in a globular or spherical structure. This class of proteins includes regulatory proteins and enzymes. Hemoglobin serves as the best example to understand the quaternary structure of proteins, which consists of four polypeptide chains, *viz.*, two α chains and two beta chains. It is a multimeric (hetero-tetramer) protein and serves the best example of allosteric proteins. It consists of 04 porphyrin rings and Fe^{2+} cofactors for carrying oxygen molecules.

6. Proteome profiling

Protein profiling or expression profiling of protein signature is important to know the status of proteins in a cell, tissue, and organism. The basic aim of protein profiling is to have insights into the physiological state and diagnose diseases and precise treatment of diverse ailments. An enormous number of proteomic techniques have been reported to investigate the proteomes that involve the use of gel-based techniques such as 2D gel electrophoresis, mass spectrometry, and tandem mass spectrometry. In addition to off-gel techniques such as isotope-coded affinity tags and multidimensional protein identification technique (MuDPIT), several spectral counting methods deliver information on proteins in a relative abundance basis (Zhang et al., 2015). Moreover, the HPLC (High Performace Liquid Chromatography)-based sample fractionation is also recommended for analyzing complex samples to enhance sequence coverage and detection at vibrant range (Chandramouli and Qian, 2009).

7. Conclusion

The investigation of eukaryotic genomes is still underway, even though most of the important eukaryotic genomes have been sequenced and analyzed to have insights into structural and functional aspects of DNA sequences. The eukaryotic genomes are constructed with both coding and noncoding sequences. It is the very low percentage of gene sequences, which repeatedly instigates the scientists now gems are hidden in about 97% of the genome in humans and a large part of noncoding sequences in other organisms. The in-depth and updated information provides a great platform to further elaborate our understanding of protein profiles through a diverse range of proteomic techniques.

References

Beagrie, R. A., Scialdone, A., Schueler, M., Kraemer, D. C. A., Chotalia, M., Xie, S. Q., Barbieri, M., de Santiago, I., Lavitas, L.-M., Branco, M. R., et al. (2017). Complex multi-enhancer contacts captured by genome architecture mapping. *Nature, 543,* 519−524.

Chandramouli, K., & Qian, P. Y. (2009). Proteomics: Challenges, techniques and possibilities to overcome biological sample complexity. *Hum Genomics Proteomics, 2009,* 239204.

Cavalli, G., & Heard, E. (2019). Advances in epigenetics link genetics to the environment and disease. *Nature, 571,* 489−499.

Nathan, D., Sterner, D. E., & Berger, S. L. (2003). Histone modifications: Now summoning sumoylation. *Proceedings of the National Academy of*

Sciences Nov,   100(23),   13118−13120.   https://doi.org/10.1073/pnas.2436173100

Dekker, J., & Mirny, L. (2016). The 3D genome as moderator of chromosomal communication. *Cell, 164,* 1110−1121.

Eytan, E., Ganoth, D., Armon, T., & Hershko, A. (1989). ATP-dependent incorporation of 20S protease into the 26S complex that degrades proteins conjugated to ubiquitin. *Proceedings of the National Academy of Sciences of the United States of America, 86*(20), 7751−7755.

Hayakawa, T., & Nakayama, J. (2011). *Physiological roles of Class I HDAC complex and histone demethylase.*

Jürgen Dohmen, R. (2004). SUMO protein modification. *Biochimica et Biophysica Acta (BBA)—Molecular Cell Research, 1695*(1−3), 113−131.

Krijger, P. H. L., & de Laat, W. (2016). Regulation of disease-associated gene expression in the 3D genome. *Nature Reviews Molecular Cell Biology, 17,* 771−782.

Lieberman-Aiden, E., van Berkum, N. L., Williams, L., Imakaev, M., Ragoczy, T., Telling, A., Amit, I., Lajoie, B. R., Sabo, P. J., Dorschner, M. O., et al. (2009). Comprehensive mapping of long-range interactions reveals folding principles of the human genome. *Science, 326,* 289−293.

Lodish, H. F. (2016). *Molecular cell biology.* New York: W.H. Freeman and Co.

Luger, K., Mäder, A. W., Richmond, R. K., & Sargent, D. F. (September 18, 1997). Richmond TJ Crystal structure of the nucleosome core particle at 2.8 A resolution. *Nature, 389*(6648), 251−260.

Man, P. Y., et al. (2003). The epidemiology of Leber hereditary optic neuropathy in the North East of England. *American Journal of Human Genetics, 72,* 333−339.

Rao, S. S. P., et al. (2014). A 3D map of the human genome at kilobase resolution reveals principles of chromatin looping. *Cell, 159,* 1665−1680.

Strahl, B. D., & Allis, C. D. (2000). *The language of covalent histone modifications.*

Su, J. H., Zheng, P., Kinrot, S. S., Bintu, B., & Zhuang, X. (September 17, 2020). Genome-scale imaging of the 3D organization and transcriptional activity of chromatin. *Cell, 182*(6):1641−1659.e26. https://doi.org/10.1016/j.cell.2020.07.032. Epub 2020 Aug 20. PMID: 32822575; PMCID: PMC7851072.

Swatek, K. N., & Komander, D. (2016). Ubiquitin modifications. *Cell Research, 26*(4), 399−422.

Twyman, R. M. (2013). *Principles of proteomics*. Abingdon: Garland Science.

Yu, M., & Ren, B. (2017). The three-dimensional organization of mammalian genomes. *Annual Review of Cell and Developmental Biology, 33*, 265–289.

Zhang, Y., et al. (2015). Tissue-based proteogenomics reveals that human testis endows plentiful missing proteins. *Journal of Proteome Research, 14*(9), 3583–3594.

Genome mapping

1. Introduction

Genome mapping is defined as the assignment/placement of a certain gene to a specific region of a chromosome, as well as the determination of the position and relative distances between genes on the chromosome. Genetic mapping is a technique that uses recombination frequencies to describe the organization of genes and their relative distances on a chromosome. Because the genes function as markers in this mapping and the maps are population specific. As a result, population mapping becomes an essential aspect of genetic mapping. Furthermore, comparing the genes to each other during genetic mapping aids in determining the order of the genes on the chromosome. A genomic map aids scientists in their exploration of the genome. A genomic map, like a road map or other recognized map, is a collection of markers that tells individuals where they are and helps them reach where they want to go. A genome map's markers might comprise short DNA sequences, regulatory regions that switch genes on and off, and genes themselves. Genome maps are frequently used to assist scientists in discovering novel genes (National Research Council, 1988). There are two kinds of maps (Fig. 2.1).

1.1 Linkage maps

Explain how genes and genetic markers are arranged throughout the chromosomes based on the frequency with which they are inherited together.

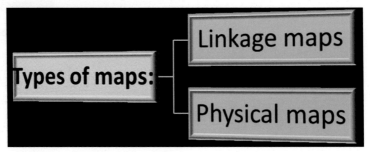

FIGURE 2.1

Types of maps.

1.2 Physical maps

Depict chromosomes and offer actual distances between chromosomal markers in nucleotide bases.

2. Markers

Any stable and hereditary variation that is measurable or detectable by an appropriate technology that can therefore be used to determine the presence of a given genotype or phenotype is classified as a molecular marker. For example, restriction fragment length polymorphisms (RFLPs), simple sequence length polymorphisms, and single nucleotide polymorphisms (SNPs) (Fig. 2.2).

Genes were the first-ever markers in the genetic mapping of species such as fruit flies in the early decades of the twentieth century. The gene, which is a segment of DNA, is essentially an abstract entity responsible for the transmission of heritable qualities from parent to

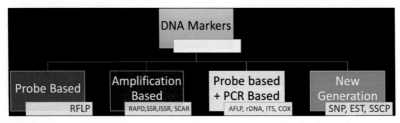

FIGURE 2.2

Different types of DNA-based markers.

child. Furthermore, each gene has at least two different forms known as alleles, which result in distinct phenotypes. These characteristics also functioned as visual markers, revealing the sites of genes for the body color, eye color, wing form, and other traits in the first fruit-fly map. Later, however, genetic mapping was based on biochemical characteristics such as blood typing. Other DNA sequence properties are also relevant for bigger genomes, such as those of vertebrates and flowering plants (Grover & Sharma, 2016).

2.1 Simple-sequence repeats

Simple-sequence repeats (SSRs), also known as microsatellites, are DNA lengths made up of short, tandemly repeated di-, tri-, tetra-, or pentanucleotide patterns. Simple sequence repeats have been discovered in all eukaryotic organisms that have been studied for them (Tautz & Renz, 1984). SSRs were discovered and developed in humans by Litt and Luty (1989) and Edwards et al. (1991) and were first used in plants by Akkaya et al. (1992). To amplify SSRs using polymerase chain reaction (PCR), unique flanking DNA sequences must be known to create primers. Electrophoresis is used to separate the amplification products by size, and silver staining or fluorescent dyes are used to view them. Because of allelic heterogeneity in the number of repeat motifs in the microsatellite, amplicons from different genotypes typically display length polymorphisms. Because SSRs are codominant markers, they may discriminate between heterozygotes and homozygotes. SSRs' key benefits are their high level of polymorphism and dependability. Many researchers have used SSRs to achieve diverse purposes (Vieira et al., 2016).

Characteristics: SSRs are single-locus markers with $1-10$ alleles in each locus; hence, SSRs are highly polymorphic.

Genotyping: Each band represents an allele with a different size in base pairs.

Genotyping technology: Particular primers surrounding a simple repetition of $1-5$ nucleotides amplify specific loci during PCR. Primers are created by screening genomic libraries using probes made up of different combinations of $1-5$ nucleotides (when the genome sequence is available, the sequence for microsatellites and synthesized PCR primers homologous to the flanking sequences can

Table 2.1 Applications and limitations of SSR.

Applications	Limitations
Highly reproducible	Not well examined
High polymorphism	Cannot be suitable across species
Multiple alleles	Sequence information needed
Genomic abundance high	
No radioactive labeling	

be screened). The number of SSR primers available in diverse plant species ranges from a few to several hundred (the SSR human map consists of about 20,000 SSRs loci). During this PCR, the primers are radioactively or fluorescently tagged. On a sequencing acrylamide gel, the amplified fragments are size segregated. A DNA sequencer may also be used to create and test the SSR banding pattern (Vieira et al., 2016).

Source of polymorphism: SSRs are one of the numerous sequence variants known as a variable number of tandem repeats, and they differ from microsatellites by the size of the core sequence (few nucleotides vs. several tens). As a result, polymorphism is defined as the number of tandem repetitions of a certain microsatellite at a specific locus (Table 2.1).

2.2 Inter simple sequence repeats

Inter simple sequence repeats (ISSRs) are DNA segments ranging in size from 100 to 3000 bp that are found between neighboring, oppositely orientated microsatellite sections. ISSRs are amplified using PCR using microsatellite core sequences as primers and a few chosen nucleotides as anchors into nonrepeat neighboring areas (16–18 bp). Approximately 10–60 fragments from various loci are created concurrently, separated by gel electrophoresis, and scored as the presence or absence of fragments of a certain size. ISSR analysis techniques include single primer amplification reaction, which uses a single primer having just a microsatellite's core motif, and directed amplification of minisatellite-region DNA, which utilizes a single primer containing only a minisatellite's core motif (Bornet & Branchard, 2001).

2.2.1 Advantage

The fundamental advantage of ISSRs is that no sequence data are required for primer synthesis. Because the techniques incorporate PCR, very little amounts of template DNA are required. Furthermore, ISSRs are dispersed at random across the genome. ISSR analysis may be used in investigations involving genetic identity, parenthood, clone, and strain identification, and taxonomy analyses of closely related species due to the multilocus fingerprinting patterns acquired. ISSRs are also thought to be effective in gene mapping investigations.

2.2.2 Disadvantage

Because ISSR is a multilocus method, one downside is that comparable sized pieces may not be homologous. Furthermore, ISSRs, like randomly amplified polymorphic DNAs (RAPDs), might have repeatability issues (Fig. 2.3).

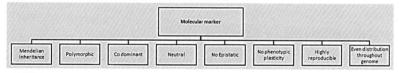

FIGURE 2.3

Properties of good molecular marker.

2.3 Randomly amplified polymorphic DNA

RAPDs are random sequence DNA fragments amplified by PCR using short synthetic primers (generally 10 bp). These oligonucleotides act as both forward and reverse primers and can typically amplify fragments from 1 to 10 genomic locations at the same time. Amplified fragments, typically in the 0.5—5 kb size range, are separated by agarose gel electrophoresis, and polymorphisms are identified as the presence or lack of bands of certain sizes following ethidium bromide staining. These polymorphisms are thought to be caused mostly

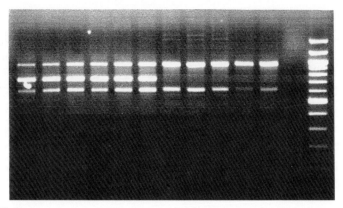

FIGURE 2.4

Representative gel pic of RAPD marker.

by variations in primer annealing sites, although they can also be caused by length changes in the amplified sequence between primer annealing sites (Fig. 2.4) (Ajmone-Marsan et al., 2002; Hadrys et al., 1992).

2.3.1 Advantages and applications

The fundamental benefit of RAPDs is that they are rapid and simple to test. Because PCR is used, very little amounts of template DNA are required. Because random primers are commercially accessible, no sequence data are required for primer synthesis. Furthermore, RAPDs have a very high genomic abundance and are scattered randomly across the genome. RAPDs have been utilized for a variety of applications, ranging from individual-level research (e.g., genetic identification) to studies involving closely related species. RAPDs have also been used in gene mapping research to fill gaps that other markers do not cover. Arbitrarily primed polymerase chain reaction (AP-PCR), which utilizes longer arbitrary primers than RAPDs, and DNA Amplification Fingerprinting, which employs shorter, 5—8 bp primers to yield a greater number of fragments, are both variations of the RAPD approach. Multiple arbitrary amplicon profiling refers to approaches that use a single arbitrary primer (Hadrys et al., 1992).

2.3.2 Limitations

Almost all RAPD markers are dominant, which means that it is impossible to tell whether a DNA segment is amplified from a locus that is heterozygous (one copy) or homozygous (two copies).

- Because PCR is an enzymatic process, the quality and concentration of template DNA, quantities of PCR components, and PCR cycle conditions can all have a significant impact on the result. As a result, the RAPD approach is famously laboratory dependent and requires meticulously defined laboratory protocols to be repeatable.
- Mismatches between the primer and the template can result in both the complete lack of PCR product and a just reduced quantity of the product. As a result, the RAPD findings might be difficult to decipher.

2.4 Restriction fragment length polymorphism

RFLP is a divergence in nucleotide Sequence detectable by the presence of fragments of varying lengths after digestion of the DNA samples in question with appropriate restriction endonucleases. As a molecular marker, RFLP is restricted to a single clone/restriction enzyme combination. Most RFLP markers are codominant (both alleles would be detectable in a heterozygous sample) and highly locus-specific. An RFLP probe is a tagged DNA sequence that hybridizes with one or more pieces of the digested DNA sample after gel electrophoresis, revealing a distinct blotting pattern indicative of a certain genotype at a specific locus. As RFLP probes, short, single- or low-copy genomic DNA or cDNA clones are commonly utilized. The difference in the size of a DNA restriction fragment (restriction map) across individuals is referred to as RFLP. It can be used as a genetic marker to aid in the study and mapping of a vast genome. RFLP is based on the idea that minor differences in DNA sequence can cause restriction enzyme cutting patterns to change. A single base-pair change in a specific chromosome, for example, or small deletions or insertions of a base pair may eliminate a site for restriction enzyme action, resulting in a considerable size difference in DNA restriction fragments. The inherited size difference in RFLPs provides a large number of linkage markers for tracking defective genes through populations (Ajmone-Marsan et al., 2002; Marsh, 1999).

In genome mapping and variation analysis, RFLP probes are extensively utilized (genotyping, forensics, paternity tests, hereditary disease diagnostics, etc.).

2.4.1 Application

The RFLP method was used in early approaches such as genetic fingerprinting, identifying items recovered from crime scenes, determining paternity, and analyzing genetic diversity or breeding trends in animal populations. RFLPs may also be used to study diversity and phylogeny, ranging from individuals within populations or species to closely related species. RFLP variation analysis in genomes, according to Al-Samarai & Al-Kazaz (2015) is a valuable tool for genetic disease research and genome mapping. The chromosomal location of a certain disease gene is determined by analyzing the DNA of disease-affected family members and looking for RFLP alleles that have an inheritance pattern similar to the condition. RFLPs have been widely used in gene mapping studies due to the prevalence of different restriction enzymes and their random distribution across the genome. It is used to examine unique patterns in DNA fragments to genetically distinguish between species.

2.5 Amplified fragment length polymorphism

Amplified fragment length polymorphism (AFLP) is a PCR-based approach that generates and compares unique fingerprints for genomes of interest by selectively amplification a selection of digested DNA fragments. The strength of this approach stems primarily from the fact that it does not require prior knowledge of the targeted genome, as well as its great repeatability and sensitivity for identifying polymorphism at the DNA sequence level (Mueller & Wolfenbarger, 1999).

Vos et al. (1995) initially described AFLP, a PCR-based finger-printing technology. Several modified methods have been recorded since then, although they usually comprise five key steps:

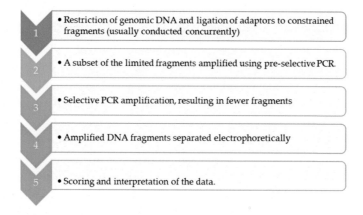

1. • Restriction of genomic DNA and ligation of adaptors to constrained fragments (usually conducted concurrently)

2. • A subset of the limited fragments amplified using pre-selective PCR.

3. • Selective PCR amplification, resulting in fewer fragments

4. • Amplified DNA fragments separated electrophoretically

5. • Scoring and interpretation of the data.

The strength of AFLP analysis stems from its capacity to create numerous marker fragments for any organism fast and without prior knowledge of genome sequence. Furthermore, AFLP uses just a tiny quantity of beginning templates and has significantly greater repeatability than other fingerprinting techniques such as RAPD and ISSR. Despite the fact that AFLP is a labor-intensive technology, it is readily multiplexed and is routinely used to amplify hundreds of genomic fragments from hundreds of individuals in the same batch.

2.6 Single nucleotide polymorphism

Single nucleotide polymorphism, or SNP (pronounced "snip"), is a variation between an individual's DNA sequence at a single place. SNP is defined as a variant in which more than 1% of a population does not carry the same nucleotide at a certain place in the DNA sequence. If an SNP exists within a gene, the gene is said to have several alleles. SNPs may cause differences in the amino acid sequence in certain circumstances. SNPs, on the other hand, are not just related to genes; they can also be found in noncoding areas of DNA. Because of their widespread frequency, accessibility of analysis, inexpensive genotyping costs, and the ability to conduct association studies using statistical and bioinformatics techniques, SNPs are regarded as the most useful biomarkers for disease diagnosis or prognosis. As a result, SNPs have recently acquired prominence as important drivers in disease-association studies. In mammals, hundreds of thousands of SNPs have been identified during the last decade to uncover connections with complicated clinical problems and phenotypic features related to hundreds of prevalent diseases (Srinivasan et al., 2016; Welter et al., 2014; Wijmenga & Zhernakova, 2018). SNPs are the most common type of sequence variation in genomes, and they are widely accepted as efficient genetic markers for revealing the evolutionary history and common genetic differences that explain heritable risk for common diseases. Genome-wide association study (GWAS) research has gained significance for discovering susceptibility genes in complex diseases. GWAS analysis is aided by information like genotype frequencies, linkage disequilibrium, and recombination rates across populations.

In plants, SNPs can also be employed to find new genes and their functions by influencing gene expression and the actions of transcriptional and translational promoters. As a result, they may be accountable for phenotypic differences between individuals in terms of improving agronomic traits. It is also critical to understand the placement of SNPs in the genome because SNPs in the coding area can have a significant impact on the activity and thermostability of an enzyme or a related product. It can also depend on the locations of the substituted amino acids because some amino acids affect the activity of the expressed regions. Recent technical improvements have made it easier to find distinct SNPs that can be used to generate new products.

The main idea underlying SNP detection is to either identify a previously unknown polymorphism or search for an already-known polymorphism. The detection approaches are classified into two categories: in vitro techniques and in silico techniques (Fig. 2.5). In in vitro, nonsequencing approaches include restriction digestion-based RFLP, CAPs, dCAPs, DNA Confirmation, SSCP, DGGE, TGGE, target-induced local legions in the genome, sequence-based approaches like locus-specific PCR amplification, whole genome short gun, overlapping regions, resequencing approaches like pyrosequencing, and Maldi-Top (Fig. 2.6).

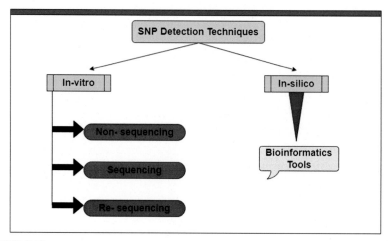

FIGURE 2.5

The techniques involved in SNP detection.

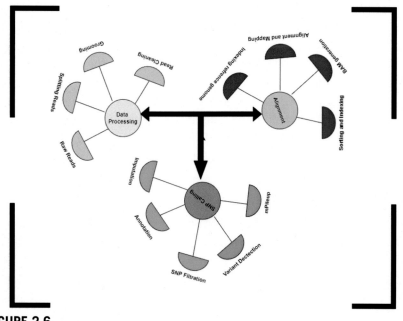

FIGURE 2.6

Important steps involved in SNP detection.

2.6.1 Advantages

- It is a codominant marker
- Most common in the genome
- Low mutation rate
- Protected for generations
- High ethnicity information and phenotype
- Data Multiplex capacity is high (>1000 SNP microarray technique)
- Fast and fully automated possible analysis

2.6.2 Disadvantages

- Low discrimination power per allele
- Lower sensitivity than short tandem repeat (STRs)
- Difficulty in the analysis of a mixed sample

3. Physical mapping

Physical mapping, on the other hand, is a technique for determining the physical distance between two genes. Physical mapping is significant owing to the poor resolution of genetic maps due to fewer crossings and their limited accuracy. It also provides the actual distance between markers using the number of nucleotides. The most common types of physical mapping approaches are restriction mapping, fluorescent in situ hybridization (FISH), and sequence-tagged sites (STS) (Deonier et al., 2005; Oslan et al., 1989).

3.1 Restriction mapping

Restriction sites are used as DNA markers in restriction mapping. There are a few polymorphic restriction sites utilized; however, there are many nonpolymorphic restriction sites used. In general, the simplest technique to create a restriction map is to compare the fragment sizes produced by two separate restriction enzymes with different target sequences. However, restriction mapping is more suitable for applications of DNA segments with few cut sites. Using uncommon cutters with few cut sites, it is still possible to examine whole genomes bigger than 50 kb. Furthermore, optical mapping is a technique for creating ordered, genome-wide, high-resolution restriction maps from single, dyed molecules of DNA.

3.2 Fluorescence in situ hybridization

The site of the marker on the chromosome may be directly seen using fluorescent in situ hybridization. It does this by the hybridization of radioactive or fluorescent probes. Also, it employs extremely condensed metaphase chromosomes. This, however, results in low-resolution mapping. As a result, using mechanically stretched metaphase or nonmetaphase chromosomes might improve resolution. FISH is the most compelling approach for detecting particular DNA sequences, diagnosing chromosomal anomalies, mapping genes, and identifying novel oncogenes or genetic abnormalities that contribute to many forms of cancer. FISH entails annealing DNA or RNA probes coupled to a fluorescent reporter molecule with a specific target sequence of sample DNA, which can then be seen via fluorescence microscopy. The methodology has since been developed to allow simultaneous screening of the entire genome utilizing multicolor whole chromosome probe methods such as multiplex FISH or spectral karyotyping, or an array-based method employing comparative genomic hybridization. FISH has dramatically transformed the area of cytogenetics and is now acknowledged as an accurate diagnostic and discovery tool in the battle against genetic conditions. In metaphase or interphase cells, FISH may detect particular regions of certain DNA sequences. This approach, which was originally developed for mammalian chromosomes, was used to plant chromosomes for the first time by Schwarzacher et al. (1989). FISH has been used to identify 18S.26SrRNA and repetitive DNA sequences in plant chromosomes such as those found in Aegilops, Hordeum, Oryza, Arabidopsis, Brassica, soybean, and barely (Shakoori, 2017).

3.2.1 Types of FISH

Since the Human Genome Project widely recognized FISH as a physical mapping technique to support massive nucleotide sequencing, it has become a more convenient and popular technique in other areas of biological and medical research such as clinical genetics, neuroscience, reproductive medicine, cellular genomics, and chromosome biology (Table 2.2).

Table 2.2 Some of the techniques described below demonstrate FISH's flexibility.

S. No	Name	Application
01	Centromere-FISH (ACM-FISH)	It is to detect chromosomal aberrations in sperm cells.
02	armFISH	It detects chromosomal abnormalities in the p- and q-arms of all 24 human chromosomes except the Y and acrocentric chromosomes.
03	Catalyzed reporter Deposition-FISH (CARD-FISH)	It is extremely effective for detecting, identifying, and quantifying microorganisms active in bioleaching processes.
04	Cellular compartment analysis of temporal (Cat) activity by fish (catFISH)	This method is used to study the interconnections of neuronal populations associated with various behaviors.
05	Cytochalasin B (CB-FISH)	Combining the traditional CB-FISH methodology with the 24-color SKY technology makes it easier to analyze the chromosomal content of micronuclei.
06	Chromosome orientation (CO)-FISH	It determines the relative orientation of two or more DNA strands along a chromosome. It is also effective in analyzing chromosomal translocations and inversions.
07	Comet-FISH	It is used to identify DNA damage in specific genomic regions. This method is now used effectively to measure telomere sensitivity to damage.
08	e-FISH	e-FISH is a FISH simulation tool that uses BLAST to anticipate the results of hybridization experiments. This application

Table 2.2 Some of the techniques described below demonstrate FISH's flexibility.—*cont'd*

S. No	Name	Application
		was created as a bioinformatics tool for determining the best genomic probes for hybridization investigations.
09	Quantitative-FISH (Q-FISH)	This approach has mostly been used to count the number of telomere repeats on a certain chromosome.
10	Rainbow-FISH	This procedure distinguishes various phylogenetic categories of bacteria by using unique 16S rRNA-targeted oligonucleotide probes.
11	Comparative Genomic Hybridization (CGH)	This approach has allowed us to comprehend those cancers of the same kind have comparable patterns of DNA gains and losses and that the frequency of alterations rises with tumor advancement.

3.2.2 Advantages

1. FISH has now become an essential method for gene mapping and chromosomal aberration characterization.
2. Chromosome abnormalities on certain chromosomes or chromosomal locations can be easily defined using differently labeled probes.
3. FISH can be used in medicine for disease diagnosis, prognosis, and remission evaluations, such as cancer.
4. The availability of molecular information in the context of cell shape is one of the great features of FISH over traditional molecular biology.
5. In addition, the FISH technique enables genome-wide screening of chromosomal gains and losses, which is comparable to in situ hybridization.

3.3 Sequence-tagged site mapping

Sequence-tagged mapping is a high-resolution, quick, and less sophisticated mapping process. As a result, it is the most powerful physical mapping approach and the one responsible for producing the most comprehensive maps of huge genomes. An STS, or sequence-tagged site, is often a short DNA sequence of 100—500 bp in length that is clearly identifiable and occurs just once in a certain chromosome or genome. As a result, an STS map may be created by combining overlapping DNA fragments from a single chromosome. On agarose or polyacrylamide gel, STS-based PCR provides a simple and repeatable pattern. STS markers are usually codominant, allowing heterozygotes to be identified from the two homozygotes. The DNA sequence of an STS may contain repeated elements and sequences seen elsewhere in the genome, but as long as the sequences at both ends of the site are unique and conserved, researchers may uniquely identify this region of the genome using technologies available in any laboratory (Williams, 1995; Oslan et al., 1989). STS markers include microsatellites (SSRs, STMS, or SSRPs), SCARs, CAPs, and ISSRs.

References

Ajmone-Marsan, P., Negrini, R., Milanesi, E., Bozzi, R., Nijman, I. J., Buntjer, J. B., … Lenstra, J. A. (2002). Genetic distances within and across cattle breeds as indicated by biallelic AFLP markers. *Animal Genetics, 33*(4), 280—286.

Akkaya, M. S., Bhagwat, A. A., & Cregan, P. B. (1992). Length polymorphisms of simple sequence repeat DNA in soybean. *Genetics, 132*(4), 1131—1139.

Al-Samarai, F. R., & Al-Kazaz, A. A. (2015). Applications of molecular markers in animal breeding. *American Journal of Applied Scientific Research, 1*, 1—5.

Bornet, B., & Branchard, M. (2001). Nonanchored inter simple sequence repeat (ISSR) markers: Reproducible and specific tools for genome fingerprinting. *Plant Molecular Biology Reporter, 19*(3), 209—215.

Deonier, R. C., Waterman, M. S., & Tavaré, S. (2005). *Physical mapping of DNA* (pp. 99—119). New York: Springer.

Edwards, A., Civitello, A., Hammond, H. A., & Caskey, C. T. (1991). DNA typing and genetic mapping with trimeric and tetrameric tandem repeats. *American Journal of Human Genetics, 49*(4), 746.

Grover, A., & Sharma, P. C. (2016). Development and use of molecular markers: Past and present. *Critical Reviews in Biotechnology, 36*(2), 290–302.

Hadrys, H., Balick, M., & Schierwater, B. (1992). Applications of random amplified polymorphic DNA (RAPD) in molecular ecology. *Molecular Ecology, 1*(1), 55–63.

Litt, M., & Luty, J. A. (1989). A hypervariable microsatellite revealed by in vitro amplification of a dinucleotide repeat within the cardiac muscle actin gene. *American Journal of Human Genetics, 44*(3), 397.

Marsh, T. L. (1999). Terminal restriction fragment length polymorphism (T-RFLP): An emerging method for characterizing diversity among homologous populations of amplification products. *Current Opinion in Microbiology, 2*(3), 323–327.

Mueller, U. G., & Wolfenbarger, L. L. (1999). AFLP genotyping and fingerprinting. *Trends in Ecology & Evolution, 14*(10), 389–394.

National Research Council. (1988). *Mapping and sequencing the human genome.*

Olson, M., Hood, L., Cantor, C., & Botstein, D. (1989). A common language for physical mapping of the human genome. *Science, 245*(4925), 1434–1435.

Schwarzacher, T., Leitch, A. R., Bennett, M. D., & Heslop-Harrison, J. S. (1989). In situ localization of parental genomes in a wide hybrid. *Annals of Botany, 64*(3), 315–324.

Shakoori, A. R. (2017). Fluorescence in situ hybridization (FISH) and its applications. In *Chromosome structure and aberrations* (pp. 343–367). New Delhi: Springer.

Srinivasan, S., Clements, J. A., & Batra, J. (2016). Single nucleotide polymorphisms in clinics: Fantasy or reality for cancer? *Critical Reviews in Clinical Laboratory Sciences, 53*, 29–39. https://doi.org/10.3109/10408363.2015.1075469

Tautz, D., & Renz, M. (1984). Simple sequences are ubiquitous repetitive components of eukaryotic genomes. *Nucleic Acids Research, 12*(10), 4127–4138.

Vieira, M. L. C., Santini, L., Diniz, A. L., & Munhoz, C. D. F. (2016). Microsatellite markers: What they mean and why they are so useful. *Genetics and Molecular Biology, 39*, 312–328.

Vos, P., Hogers, R., Bleeker, M., Reijans, M., Lee, T. V. D., Hornes, M., … Zabeau, M. (1995). AFLP: A new technique for DNA fingerprinting. *Nucleic Acids Research, 23*(21), 4407−4414.

Welter, D., MacArthur, J., Morales, J., Burdett, T., Hall, P., Junkins, H., et al. (2014). The NHGRI GWAS Catalog, a curated resource of SNP-trait associations. *Nucleic Acids Research, 42*, 1001−1006. https://doi.org/10.1093/nar/gkt1229

Wijmenga, C., & Zhernakova, A. (2018). The importance of cohort studies in the post-GWAS era. *Nature Genetics, 50*, 322−328. https://doi.org/10.1038/s41588-018-0066-3

Williams, B. D. (1995). Genetic mapping with polymorphic sequence-tagged sites. *Methods in Cell Biology, 48*, 81−96.

Analysis of genomes—I

1. Introduction

1.1 Genome sequencing

Genome sequencing refers to sequencing the entire genome of an organism, instead of sequencing it gene by gene. Sequencing the entire genome will provide a wealth of data that can be studied completely. Studying the functions of genes one by one is time consuming, it leads to redundant work by many scientists, slow progress, and only incomplete information is obtained. Due to the advances in DNA sequencing technology, genome sequencing has been made possible (Brown, 2018). Genome sequencing is easy with prokaryotic organisms because their genome is small and contains very few or no repetitive sequences. But sequencing the eukaryotic genome is difficult due to its large size and the large number of repetitive sequences.

However, improvements in sequencing technology and computerized data handling have paved the way for the sequencing of even very large genomes like the human genome. After the human genome project was initiated, many high throughput automated sequencing techniques were developed and executed. Nucleotide sequence is the highest-resolution map of a genome. It provides comprehensive information about genes and their related regulatory elements and other features of a genome. The ultimate objective of a genome project is the complete DNA sequence for the organism being studied. This chapter describes the techniques and research strategies that are used during the sequencing phase of a genome project, when this ultimate objective is being directly addressed. Techniques for sequencing DNA are clearly of central importance in this context, and we begin the chapter with a detailed examination of sequencing methodology. This methodology is of little value, however, unless the short sequences that result from individual sequencing experiments can be linked together in the correct order to give the master sequences of the chromosomes that make up the genome. The second part of this chapter, therefore, describes the strategies used to ensure that the master sequences are assembled correctly.

Over the years, a number of different methods for DNA sequencing have been developed, and others are likely to become important in the future. The techniques in use today can be divided into two categories:

- Frist generation methods used in human genome project (HGP) for instance the chain-termination method, developed by Fred Sanger and colleagues in the mid-1970s.
- Compared to first-generation techniques, the massively parallel DNA sequencing approaches led to the development of next-generation sequencing (NGS) that provides a higher magnitude of sequencing and higher order accuracy and precision.

2. Sangers sequencing/chain-termination sequencing method

This method is also known as chain-termination sequencing developed by Frederick Sanger in 1977. The genome sequencing projects such as human, archaeal, and bacterial genomes were carried out before mid of the 2000s were carried by using the Sangers sequencing method. Even in current times, the genome sequences are carried out by using next-generation sequencing technologies. At the same time, the chain-termination sequencing method is still sued for sequencing the short DNA sequences such as that of Polymerase Chain Reaction (PCR) products or cloned fragments.

2.1 The basic principle of chain-termination DNA sequencing

It is based that single-stranded DNA differing in a single nucleotide can be resolved by using polyacrylamide gel electrophoresis. The electrophoresis is carried in a 50—80 cm capillary tube, which is feasible to resolve DNA fragments of up to 1500 nucleotides. This method involves the use of DNA polymerases used to make copies of DNA fragments to be sequenced and resolved by electrophoresis. The steps include annealing the short oligonucleotide, as a primer to the DNA template. In addition four nucleotides, viz., dNTPs: dATP, dCTP, dGTP, and dTTP act as substrates to aid addition to the growing polypeptide chain. The modifications involve the addition of differentially labeled (labeled with fluorescent molecules) four dideoxynucleotide triphosphates (ddNTPs), viz., ddATP, ddCTP, ddGTP, and ddTTP. Once the dideoxynucleotide is incorporated the elongation of the nucleotide chain is blocked due to the lack of OH group, critical for nucleophilic attack of incoming NTP to form phosphodiester bond. Due to the presence of a large quantity of normal deoxynucleotides in comparison to dideoxynucleotides the strand synthesis terminates after incorporating a large number of nucleotides. The outcome of the process is the yield of differential lengths of nucleotide chains terminated after incorporating dideoxynucleotides. The mixture is subjected to the polyacrylamide gel electrophoresis and separated according to the length.

The sequencing enzymes used in DNA sequencing must have the following salient features:

- Low $5' \rightarrow 3'$ exonuclease activity—this activity may alter the length of polymerized polynucleotide chain from $5'$ end.
- Low $3' \rightarrow 5'$ exonuclease activity—this activity may alter the length of polymerized polynucleotide chain from $3'$ end.
- High processivity—It will not be disassociated during the incorporation of nucleotides except after incorporating dideoxynucleotide.

The above qualities were largely accomplished by using Klenow polymerase. The only limitation observed in the case of Klenow polymerase is that it naturally terminates after incorporating 250 bases of nucleotides. This limitation is replaced by using *Taq* polymerase instead of Klenow polymerase. Later has no exonuclease activity and a higher level of polymerase activity.

3. Whole-genome shotgun sequencing

It is very tedious to sequence complex eukaryotic genomes as they have a huge number of repetitive sequences. The basic workflow involves the overlapping of DNA fragments in parallel and then assembling the small fragments in a computer to form large fragments in the form of contigs (Weber & Myers, 1993). For highly repetitive DNA sequences, hybrid Whole Genome Sequencing (WGS) is usually employed by cleaving the genomes into overlapping clones that are physically mapped, and subsequently, the intermediate segments are sequenced by a shotgun approach.

This technique holds a lot of critical advantages in comparison to conventional clone-by-clone sequencing. For optimization of polymorphism detection, the genomic DNA must be ideally sequenced from a wide range of individuals with differing geographic ancestries and habitats. The error during sequencing may be up to 1% and the rate of polymorphism may be on the order of 0/1%. For instance, it helps in elucidating the huge number of DNA polymorphisms with more precision and speed. Through this approach, genes can be mapped through linkage analysis (Terwilliger & Ott, 1994), such that the detection of diseases is easy and precise (Weber, 1994). Likewise, WGS also aids in the detection of submicroscopic chromosomal rearrangements (Lupski et al., 1991), to find paternity and forensic testing (Smith, 1995; Urquhart et al., 1995), to study evolutionary trends (Jorde et al., 1995), population biology and recombination (Morell et al., 1995; Weber et al., 1993). In addition, studying the nucleotide polymorphism in regulatory sequences, viz., promoter regions, etc. helps to detect the diseases.

4. Hierarchical shotgun sequencing

We have read in the last section that the shotgun sequencing technique can be applied to genomes with very less repetitive DNA sequences and it is evident that the assembly problem is straightforward, as the overlapping sequences can be merged together without an error of misassembly. Shotgun sequencing is applied to sequencing simple genomes such as viruses, bacteria, organelle genomes, and plasmids. But a greater challenge arises when tackling complex genomes containing a high number of repetitive DNA sequences that may lead to misassembly when sequencing by using a shotgun approach. To solve this problem, scientists modified the shotgun approach to Hierarchical shotgun sequencing. This sequencing method is a modification of the shotgun sequencing approach and it aided in sequencing the human genome. This method is also used for sequencing genomes of plants such as wheat, barley, and other plants. The method involves the presequencing phase during which large fragments of about 300 kb are obtained by cleaving the genomes and subsequently the larger fragments are cloned by using bacterial artificial chromosomes (BACs). The overlapping fragments of DNA are identified to form clone contigs or contiguous series. Finally, the sequence of each clone insert is then subjected to a shotgun method for assembly. This step aids in the build-up of the master sequence. The errors due to the presence of repetitive sequences can be rectified by examination of sequences of clones that overlap with the one containing the repeats.

5. Next-generation sequencing methods

The term is used to catch all terms used for describing a wide range of modern sequencing technologies that allow sequencing both DNA and RNA more quickly and cheaply than first-generation techniques. The major advantages of using NGS techniques include the following:

- No need for any prior knowledge regarding genomes or transcriptomes
- It offers to detect the various features of genes such as allelic variants, SNPs, alternatively spliced transcripts, and mutations.
- Requires nanograms of DNA or RNA volumes for sequencing
- Results obtained are highly reproducible

5.1 Major NGS techniques include

✔ **The Illumina (Solexa) sequencing:**
It works by identifying the DNA bases as each base emits a unique fluorescent signal and subsequent addition to the nucleic acid chain.

↙ **The Roche 454 sequencing:**
The technique is based on the principle of pyrosequencing that involves the detection of released pyrophosphates by light signals after the incorporation of nucleotides by polymerases in newly synthesized DNA molecules.

↙ **The ion torrent: proton/PGM sequencing:**
It measures the direct release of H+ (protons) upon the incorporation of each nucleotide by DNA polymerase.

Almost all of these methods are used to sequence the millions of DNA fragments in parallel accomplished in a single experiment. It enables the sequencing of a vast amount of data in a small number of runs at higher speed and precision. Cost saving is significant due to the sequencing of hundreds of DNA sequences in a single run or very few runs. Sequencing library preparation is common to all NGS methods. In this method, the genomic DNA is fragmented into 100−500 bps by using the sonication method (the technique involves the use of high-frequency sound waves to induce random cuts in DNA molecules). The fragments are immobilized on a solid support such as on a glass slide coated with short oligonucleotides. The DNA fragments contain short pieces of dsDNA ligated at one end as adaptors, whose sequences match the immobilized oligonucleotides.

The dsDNA-ligated DNA fragments are then denatured resulting in ssDNA stranded molecules attached to the glass slide by base pairing between their adaptor sequences and the immobilized oligonucleotides. Another set of immobilization involves the emulsion in which each bead in an aqueous droplet is transferred to the different well in a multiple array on a plastic strip. So this case involves solid support provided by metallic beads coated with streptavidin. In this method, the DNA fragments are ligated to biotin-linked adaptors, having a strong affinity with streptavidin. The strategy is aimed to attach the beads containing streptavidin with the DNA fragments ligated with biotin-adaptor molecules. These beads are then mixed and shaken in an oil−water mixture to yield the emulsion. The final setup consists of PCR amplification of DNA molecules attached to the beads. Additionally, adaptors serve as primer binding sites for amplification through PCR. In this case, PCR is operated in an oil emulsion such that PCR products are retained in each droplet before deposition on the plastic strip.

5.2 The sequencing methods included in next-generation sequencing

The evolution of NGS methods is grossly based on decreasing the cost of sequencing and the efficacy of the method. The most popular method currently employed includes the reversible terminator sequencing method. This method involves the use of modified nucleotides that terminate the DNA synthesis. In this method, the termination step is reversible, as the chemical group attached to the 3′-carbon of the modified nucleotide can be removed. The blocking chemical group is usually a fluorescent label different for each nucleotide. The reaction mixture does not include normal deoxynucleotides. The label is then removed by enzymes enabling the addition of next terminator nucleotide. The process is initiated by primer that anneals with the adapters ligated to the ends of DNA fragments during the process of library preparation. All the clusters in the library are sequenced at the same time. This method generates short sequence reads with a maximum length of up to 300 bp. The technique is generally referred to as Illumina sequencing. The name is based on the sequencing instrument supplier company.

Before Illumina sequencing, the most common method used for sequencing was pyrosequencing. This method relied on the use of a reaction mixture containing deoxynucleotides, and there is no termination during the copying of the template. The chemiluminescence generated by sulfurylase from the released pyrophosphate during the addition of deoxynucleotide to the 3′-end of the growing strand. Subsequently, the chemiluminescence signals copying of each nucleotide during the process. The deoxynucleotides are therefore added individually in a repetitive series (e.g., A, then T, then G, then C, then A, then T, etc.) and the pattern of light emissions is used to deduce the order in which nucleotides are incorporated into the growing strand.

In this context, the pyrosequencing of immobilized fragment libraries on magnetic beads has been used in the 454 sequencing NGS method. This method can give reads of up to 1000 bp long polynucleotide chain, which is enough to generate 700 Mb DNA sequence per run.

The ion torrent method uses a prototypical method by using a pyro-sequencing approach with a repetitive series of nucleotides retained from immobilized libraries. Though, in this approach, the detection system is directed at the H^+ ions that, along with pyrophosphate, are released every time a nucleotide is added into the polymerizing strand. In the ion torrent method, the reactions are carried out by immobilizing the DNA fragments on beads made of acrylamide. In which each bead is lined with an ion-sensitive field effect transistor (ISFET).

The electronic pulse is generated by ISFET which detects the hydrogen ions each time with the flow of nucleotides. Another NGS approach, called Sequencing by Oligonucleotide Ligation and Detection (SOLiD) involves a different approach. Sequences are deduced by using different complimentary oligonucleotides rather than polymerase-directed synthesis. The process is initiated by the annealing of primer to the DNA template. The primer is annealed at the adaptor site of the template. Approximately around 1024 oligonucleotides representing each of the possible five-nucleotide sequences are added including DNA ligase. One of these oligonucleotide sequences will have a complimentary sequence adjacent to the primer annealing site. The process is continued until 50–75 nucleotides of the template have been covered. Remember that each of these oligonucleotides is tagged with a fluorescent marker, at least four different types of fluorochromes. This means, about 256 oligonucleotides are tagged with fluorescent markers among 1024 added oligonucleotides. So the detection of markers attached to oligonucleotides assigns a color to the terminal pair of nucleotides. The sequencing process is repeated such that the second primer will anneal to the target site almost position offset by one nucleotide ($n - 1$ position) compared to the first primer. The process is repeated for other sets of primers, viz, primers 3, 4, and 5. Due to double reads, it is found that SOLiD is highly accurate in sequencing the DNA fragments. The SOLiD is aimed to detect polymorphisms in DNA samples from different individuals for analysis.

6. Genome sequencing projects

6.1 The human genome project

The project was a collaboration of international scientists to completely map and understand all the genes harbored in the human genome. The efforts were put to sequence about 3 billion bps found in human genomes. The project was articulated in 1998 by a special committee of the National Academy of Sciences, US and adapted through 5-year plan in joint collaboration with Department of Energy (DoE) and the National Institutes of Health (NIH). James D. Watson was the lead scientist and Director of the HGP. In 1990, a joint research plan was completed in form of a publication titled "Understanding Our Genetic Inheritance: The Human Genome Project, The First Five Years, FY 1991−95." Even though the HGP timeline was extended to 15 years from the year 1990.

The major ways or order of deciphering the HGP were as follows:

1. To determine the order or sequences of the nucleotide bases in human genome
2. For locating the genes, maps were designed for majority of regions in chromosomes
3. Production of linkage maps for tracking inherited traits.

Human genome project revealed humans have probably 20,500 genes. In addition, the project helped in providing detailed information regarding structure, organization, and functional aspects of almost all the genes of humans. The first draft of the human genome was published by International Human Genome Sequencing Consortium in Nature journal in February 2001 comprising up to 90% of the human genome accounting for up to 3 billion bps. The consortium was led by more than 2800 researchers from different countries. The finding opened new windows of facts viz., estimated number of genes, that is, 50,000 to as many as 140,000. The full sequence was further completed in April 2003. The sequencing of human genome opened the window for sequences genomes of organisms such as mice, flatworms, and fruit flies.

The HGP was completed in the following phases:

The first phase also called as shotgun phase, in which the human chromosomes are divided into DNA segments of equal size, which were further divided into smaller fragments for sequencing. The second phase also called as finishing phase just operated to fill the gaps between the partially resolved sequences in the first phase.

The shotgun phase results in sequences of up to 90% of the human genome. The phase was accomplished in three major steps (Fig. 3.1):

- Step 1: To obtain a DNA clone for sequences
- Step 2: The second one is to sequence the clones obtained in the first step
- Step 3: The sequence assembly from multiple clones for determination of overlaps to form a contiguous sequence.

The use of the hierarchical shotgun method was used to systematically generate the overlaps of individual clones of chromosomes. The clones were actually derived from DNA libraries (generated from ligated DNA fragments). The DNA fragments were mapped to chromosomal regions by STS (sequence-tagged sites) screening with less than 500 bps. The library clones were digested with restriction

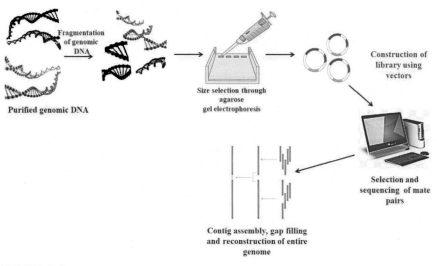

FIGURE 3.1

Major steps involved in whole-genome shotgun sequencing approach used for construction of human other genomes.

enzyme HindIII (AAGCTT) and the resulting fragments were subjected to agarose gel electrophoresis for size determination. Each of the libraries exhibited a range of DNA fragments visualized as "fingerprint," used for comparing the other fingerprints to obtain complete contagious sequences. Other methods such as FISH (fluorescence in situ hybridization) have been used to map libraries specific to chromosomal regions. Collectively, DNA fingerprints, STS, and FISH data help in generating contigs consisting of BAC library clones representing 24 different human chromosomes. The individual BAC clones were further fragmented to generate smaller fragments and then subsequently cloned in vectors to generate BAC-derived shotgun libraries.

Before completing the first phase of genome sequencing through HGP, another Biotechnology company named as Celera Genomics also entered the race to sequence the human genome. The team was led by Dr. Craig Venter, who proclaimed that our team has completed sequencing of the entire human genome in 3 years. The generated two data sets of sequences, viz., the first data set consists of 27.27 million DNA sequence reads, each with an average length of 543 bps and the second data set was described from HGP data. This company then used the whole-genome assembly method and the regional chromosome assembly method to sequence the entire human genome. The outline of the whole-genome sequencing method comprises BAC fragments assembly to produce the consensus sequences and contigs. The contigs are then connected into scaffolds by pairing the end sequences. The scaffolds are then mapped to the genome using STS information. In 2001 February, both groups published the findings of the sequenced human genome in separate articles.

HGP has resulted in identifying the single change in nucleotide sequences to trace human diseases.

6.2 Some facts of HGP

- Complete physical maps of chromosomes Y and 21 and detailed RFLP maps of the X chromosome and all 22 chromosomes were published in 1992.
- Announcement of Celera Genomics in 1998 by J. Creig Ventor, goal to sequence the human genome in 3 years (Venter, Smith, & Hood, 1996).

- In December 1999, the first draft sequence of an entire human chromosome 22 was published.
- Draft sequence of Chromosome 21 was published a few months later on May 2000.
- On June 26, 2000, accompanied by the President of the United States, the leaders of two projects, Francis Collins and Crag Ventor, jointly announced the completion of their genome drafts, which appeared 8 months later in print media
- The 90% of the genome was sequenced mostly comprising euchromatin and the rest missing 10% comprised constitutive heterochromatin.
- The sequences of the human genome provided one surprise: there appeared to be only 25,000–30,000 genes rather than the estimated 50,000–120,000 genes suggested in earlier studies.
- Over 40% of human proteins share similarity with the Drosophila and *C. elegans.*
- On an average, there is 1 gene per 145 kb in human genome.
- The average human gene length is about 27,000 bp in length and contains 9 exons.
- Exons make up only 1.1% of the genome.
- Introns make up 24%.
- 75% intergenic DNA—out of this much % age 44% is derived from transposable genetic elements.
- Watson and Craig Ventor were the first two individual's humans whose genome was sequenced.
- Jeffery M. Kidd and coworkers mapped and sequenced structural variation in eight genomes of diverse origins. The individuals from diverse ancestries—African, Asian, and European.
- Francis and Collins have organized a consortium, ENCODE (ENCyclopedia Of DNA Elements) to identify all of the nongenic functional elements in the human genome.
- Vast noncoding sequences were called as "Dark matter."
- Recent studies indicate that about 80% of the genome is transcribed into noncoding RNA consisting of important, small regulatory miRNAs.
- In addition, dark matter (noncoding sequences) is chemically inert due to chemical modifications—epigenetics (methylation).

- In addition to the Genome consortium, another international consortium—the Human Proteome Organization to study the structure and function of the total proteome.

6.3 Arabidopsis genome

The sequencing of the Arabidopsis genome was proposed in 1989 by the Biological, Behavioral, and Social Sciences Directorate of the National Science Foundation. It has a diploid set of chromosomes with $2n = 10$ chromosomes. It was the first plant genome sequenced by scientists. This plant contains 02 chromosomes (1 and 3), which are longer than the others and two chromosomes viz., 2 and 4 having Nucleolar Organizer Region (NORs) at telomeric regions. On the basis of the position of the centromere the Arabidopsis consists of 01 metacentric (chromosome 1), 02 acrocentric (chromosomes 2 and 4), and 02 submetacentric (chromosomes 3 and 5). The genome size is ~ 120 Mb with fewer repeated DNA sequences. It encodes $\sim 27,600$ protein-coding genes and 6500 noncoding genes. The sequencing of Arabidopsis was carried out by using Bacterial Artificial Chromosome (BAC), Yeast Artificial Chromosome (YAC), P1 (bacteriophage), and transformation-competent artificial chromosome libraries. Genetic maps and physical maps were integrated to anchor clones and contigs. The final map was built using a combination of fingerprint analysis of BAC libraries.

6.4 The introduction to microbial genome sequencing

The bacteriophage lambda with a genome size of 48,502 was the largest genome sequenced by random strategy completed by Sanger et al., in 1982. Then afterword the sequencing technologies have not progressed due to a lack of sufficient advances in computational approaches. Table 3.1 provides the major breakthroughs in the sequencing of bacterial genomes.

6.4.1 Mycoplasma genitalium

This organism grows in the respiratory and genital tract of humans, and its genome size is up to 580 kbs. This organism is one of the smallest known free-living bacteria. Its genome consists of 517 genes of

Table 3.1 Sequence data of *Arabidopsis thaliana*.

Genome assemblies	113
Sequence reads	134
Median total length (Mb)	120,108
Median protein count	27,334
Median GC%	36.146

which 37 code for RNAs and 480 code for proteins. It is reported that about 90 proteins are involved in translation and 29 in DNA replication. The findings also suggest that about 121 genes are not essential for the survival of *M. genitalium* (Table 3.2).

6.4.2 Haemophilus influenza

The first bacterial genome sequenced was that of *Haemophilus influenza*. It is a Gram-negative coccobacillus. The genome consists of 1.8 MB coding approximately 1743 genes (Fleischmann et al., 1995). More than 40% of genes have unknown functions. This bacterium lacks three essential genes for the Krebs cycle (Table 3.3).

6.5 Concept of genome annotation

It is the process of identifying functional elements of a genome. Consequently, the functional as well as structural features are understood as regions contained within organized functional regions of

Table 3.2 Genomic information of *Arabidopsis thaliana*.

Type of genomic unit	Size in Mbs	Number of protein coded	Gene
Chromosome no. 1	30.43	12,653	9701
Chromosome no. 2	19.70	7599	6312
Chromosome no. 3	23.46	9474	7624
Chromosome no. 4	18.59	7426	5842
Chromosome no. 5	26.98	10,995	8419
Mitochondrial DNA	0.37	33	284
Chloroplast DNA	0.15	85	129

Table 3.3 Major breakthroughs in evolution of bacterial sequencing.

Date	Scientific achievements in bacterial systems
1997	Fred Sanger invented dideoxy chain terminator sequencing
1981	First time mitochondrial genome of humans was sequenced
1995	*H. influenzae* and *M. genitalium* genomes were first-time sequenced
1996	The genomes *M. genitalium* and *M. pneumonia* were completely sequenced
1997	Genome sequences from *B. subtilis* and *E. coli*
1998	First time *M. tuberculosis* genome was sequenced
1999	The *H. pylori* was the first species sequenced
2000	Reverse vaccinology was started from meningococcal genome sequence
2001	Genome sequences of *S. aureus* and *M. leprae*
2002	Genomes of different strains of *B. anthracis* were sequenced
2003	*T. whipplei* genome was sequenced
2004	Genome sequence of mimivirus
2005	Whole-genome sequencing of *M. tuberculosis*
2006	First next-generation sequencer: the 454 GS20 was invented
2007	Illumina Genetic analyzer 2 was launched in market
2008	The sequencing of 100 strains of *S. Typhi* was kick started
2009	Start of transposon-sequencing in Salmonella genome
2010	Metagenomics to identify human gut microbiome
2012	First breakthrough to program the CRISPR–Cas bacterial RNP for genome editing
2013	Metagenomics used to analyze *E. coli*
2014	Dense sequencing of >3000 *S. pneumoniae* isolates
2015	Genome sequencing for routine surveillance of pathogenic bacteria such as tuberculosis and salmonellosis

the genome. The sequence annotation is the protocol used to determine the properties and location of sequences. The different stages of genome annotation are briefly mentioned below:

- **The ab initio and similarity**
  This method is employed to identify the protein-coding genes using signals that characterize and define the gene structure throughout

the genome. The method predicts the genes coding for proteins by similarity.

- **Ab initio gene prediction**

  The initiation and termination codons in transcription help the computational methods to identify the eukaryotic and prokaryotic genes coding for proteins. Different tools were sued codon bias to annotate the ESTs, and mRNA fragments to annotate the genes. Two programs were used to obtain the Ab Initio Gene Prediction GENEID and GENSCAN.

- **The comparative and homology**

  Two main classes of gene identification have been used, which include the comparison of DNA query sequences of mRNA sequences or target sequences and the use of sequences from databases. In addition, it also involves the comparison of two or more genomic sequences for gene prediction.

  Other methods used for gene annotation include whole genomes, RNA, and regulatory sequence analysis, by using annotation by genome-wide RNA mapping.

7. Conclusion

The process of knowing genomes in their entirety is critical to investigate the wealth of data and is considerably obtained by genome sequencing technologies. For instance, chain termination sequencing was primarily used to sequence very short DNA segments, for example, PCR products. Whereas, NGS used a wide range of advanced techniques to sequence thousands or millions of DNA fragments in one go of experiment. For sequencing small genomes, the shotgun method was used by examination of reads that overlap. In contrast, shotgun sequencing approaches for eukaryotes required complex assembly due to their larger size. The more complex genomes were subjected to the hierarchical shotgun method by using clone contigs that consisted of a series of clones using BAC to generate genetic maps. For instance, the evolution of second-generation or NGS techniques has revolutionized the genome sequencing of living organisms. It was NGS methods that led to decipher the entire genome of the giant panda.

References

Brown, T. A. (2018). *Genomes 4* (4th ed.). https://doi.org/10.1201/9781315226828 Garland Science.

Fleischmann, R. D., Adams, M. D., White, O., et al. (1995). Whole-genome random sequencing and assembly of *Haemophilus influenzae* Rd. *Science, 269*, 496–512.

Jorde, L. B., Bamshad, M. J., Watkins, W. S., Zenger, R., Fraley, A. E., Krakowiak, P. A., Carpenter, K. D., Soodyall, H., Jenkins, T., & Rogers, A. R. (1995). Origins and affinities of modern humans: A comparison of mitochondrial and nuclear genetic data. *American Journal of Human Genetics, 57*, 523–538.

Lupski, J. R., de Oca-Luna, R. M., Slaugenhaupt, S., Pentao, L., Guzzetta, V., Trask, B. J., Saucedo-Cardenas, O., Barker, D. F., Killian, J. M., Garcia, C. A., Chakravarti, A., & Patel, P. I. (1991). DNA duplication associated with CharcotMarie-Tooth disease type 1A. *Cell, 66*, 219–231.

Morell, R., Liang, Y., Asher, J. H., Jr., Weber, J. L., Hinnant, J. T., Winata, S., Arhya, I. N., & Friedman, T. B. (1995). Analysis of short tandem repeat (STR) allele frequency distributions in a Balinese population. *Human Molecular Genetics, 4*, 85–91.

Smith, R. N. (1995). Accurate size comparison of short tandem repeat alleles amplified by PCR. *Biotechniques, 18*, 122–128.

Terwilliger, J. D., & Ott, J. (1994). *Handbook of human genetic linkage.* Johns Hopkins University Press.

Urquhart, A., Oldroyd, N. J., Kimpton, C. P., & Gill, P. (1995). Highly discriminating heptaplex short tandem repeat PCR system for forensic identification. *Biotechniques, 18*, 116–121.

Venter, J. C., Smith, H. O., & Hood, L. (1996). A new strategy for genome sequencing. *Nature, 381*, 364–366.

Weber, J. L. (1994). Know thy genome. *Nature Genetics, 7*, 343–344.

Weber, J. L., & Myers, W. E. (1993). Human whole-genome shotgun sequencing. *Genome Research, 7*, 401–409.

Weber, J. L., Wang, Z., Hansen, K., Stevenson, M., Kappel, C., Salzman, S., Wilkie, P. J., Keats, B., Dracopoli, N. C., Brandriff, B. F., & Olsen, A. S. (1993). Evidence for human meiotic crossover interference obtained through construction of a short tandem repeat polymorphism linkage map of chromosome 19. *American Journal of Human Genetics, 53*, 1079–1095.

Analysis of genomes—II*

---

1. Introduction

The key to the regulation of cellular functions such as gene expression, DNA replication, DNA repair mechanisms, segregation patterns, stability of chromosomes, epigenetic modifications, and cell cycle progression. Additionally, regulatory proteins binding to DNA and histone proteins to maintain the chromatin structure. Henceforth, it is vital to know the DNA—protein interactions to investigate the gene location, regulatory sequences under in vivo conditions, since in vitro conditions study the DNA—protein interactions outside the cell. Consequently, the interaction of proteins and DNA sequences is the most critical aspect in understanding the mechanism of gene expression. The vital molecular and cellular studies underpinned by DNA—protein interactions include studying DNA modifications, DNA repair, DNA packing, transcriptional regulation, transcriptional process, recombination, and tracing viral infections (Ofran et al., 2007). The proteins that interact with single- or double-stranded DNA must possess DNA binding domains possessing special motifs such as helix—loop—helix, helix—turn—helix, zinc finger motifs, and leucine zippers. These proteins may bind to specific nucleotide sequence like transcription factors or nonspecific target DNA sequences like helicase enzymes, and SSBs that bind DNA at random sites. Various technical methods, viz., DNA footprinting, gel retardation assay, and chromatin immunoprecipitation (ChIP), yeast one-

---

* Tracing evolution of eukaryotes by understanding genomes, identifying the functional regions of genomes—DNA footprinting, gel retardation assay or electrophoretic mobility shift assay, and chromatin immunoprecipitation.

Principles of Genomics and Proteomics. https://doi.org/10.1016/B978-0-323-99045-5.00003-3
Copyright © 2023 Elsevier Inc. All rights reserved.

hybrid system, DNA pull-down assay, filter-binding assay, and luciferase reporter assay, are employed to investigate the DNA—proteins interactions for functional and regulatory actions of proteins and DNA sequences. The ChIP is evolved as a widely used technique to determine the protein binding sites of transcription factors (TFs), histone proteins, and other regulatory proteins (Ren et al., 2000; Spencer et al., 2003; Weinmann & Farnham, 2002; Weinmann et al., 2001, 2002; Yan et al., 2003). This chapter is aimed to describe the principles, working, and applications of technical methods employed for studying different molecular phenomena based on DNA—protein interactions. We have described the advantages as well as disadvantages associated with each technique to have clear insights into the evolution of the explained techniques.

2. DNA footprinting/deoxyribonuclease I (DNase I) protection mapping

It is an in vitro technique used to investigate the binding of a particular protein to target DNA region. The assay used to investigate the DNA—protein interactions to locate specific DNA binding sites is a modified version of the Maxam—Gilbert sequencing technique (Galas & Schmitz, 1978). The technique is very useful in the identification of DNA regions that are bound by transcription factors or those proteins, which are involved in the initiation of transcription. In combination with other methods such as chromatin immunoprecipitation or electrophoretic mobility shift assay, it is used to delineate the critical amino acid residues of protein interacting with DNA sequences. Currently, various variations that are based on both in vitro as well as in vivo setups are designed.

The basic principle of the techniques lies in the protection of DNA segments from applied DNA cleaving agents such as DNase-I through hydrolysis due to protein interaction with the specific DNA region. In this technique, the protein-bound DNA molecules are subjected to the treatment of DNA-cleaving agents followed by the running of cleaved products in polyacrylamide gel. In comparison to treated DNA, the treated DNA—protein cleaved products will show "footprints" or "gaps" where a specific protein is bound to DNA regions. The segments of DNA visible in polyacrylamide gel are protected DNA fragments bound by proteins and are visualized by exposing dried gels to X-rays, a process known as autoradiography. The regions in the gels showing no bands corresponding to the protected DNA regions are called as DNA footprints (Fig. 4.1).

2.1 Workflow of DNA footprinting (Fig. 4.2)

Step 1 The DNA fragments containing possible protein binding sites are extracted and labeled by using polymerase chain reaction (PCR) techniques

Step 2 The labeled DNA fragments and specific proteins are subjected to cleavage in a test tube by using the following agents:

✔ **DNase I**

An enzyme that cleaves phosphodiester bonds in dsDNA. The action of DNase I is controlled by EDTA (chelating agent).

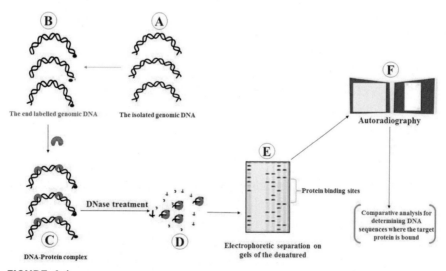

FIGURE 4.1

(A) Isolation of genomic DNA, (B) preparation of end-labeled DNA restriction fragment, (C) equilibration of the protein with DNA, (D) exposure of the equilibrium mixture to DNase I, (E) electrophoretic separation on gels of the denatured hydrolysis products, (F) autoradiography of the gel containing the DNA protected by specific protein.

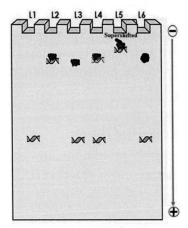

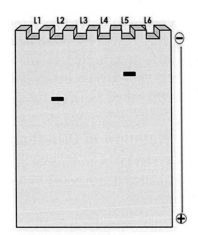

Electrophoresis of different
reaction mixtures

Detection of resolved complexes

FIGURE 4.2

Ability of protein to interact with labeled DNA by mixing the purified
protein with labeled DNA fragments and subsequently electrophoresed
in gel. The gel is then subjected to detection by autoradiography.
L-1 DNA without protein as a control
L-2 DNA—protein interaction
L-3 knotted control with protein
L-4 nonradiolabeled DNA—protein
L-5 antibody—protein—DNA interaction complex
L-6 protein and DNA not interacting and are separated individually.

✔ **Hydroxyl radicals**
  The generation of hydroxyl ions occurs in a reaction in which iron
  salts are reduced by the addition of hydrogen peroxide. They are
  capable of cleaving phosphodiester bonds in DNA fragments.
✔ **UV radiations** are also employed to fragment the DNA molecules
Step 3 This step is employed to generate a standard to compare the
outcome of our test sample. In this step, the labeled DNA fragments
are subjected to cleaving agents as mentioned earlier but without the
protein of interest. Reactions are again done in a test tube and
products act as a control for the experiment.

Step 4 The products from both the test tubes are subjected to poly-acrylamide gel and subsequently to autoradiography for developing an autoradiogram.

Step 5 The missing bands (footprints) in the gel indicate the protein binding sites in the genome

2.2 Applications of DNA footprinting

→ **In vivo footprinting:** To evaluate the protein specificity in the reference genome at many places under in vivo conditions. In this process, the immunoprecipitation by employing specific antibody DNA bound proteins of interest, followed by assessment of specific DNA regions by footprinting technical flow.

→ **Quantitative footprinting:** This aspect of footprinting helps to assess the binding strength of proteins to specific regions of DNA molecules. The strength of proteins is estimated by using different concentrations of proteins and their binding affinity to the target site.

→ **Detection by capillary electrophoresis:** In recent approaches of DNA footprinting, the labeled DNA fragments are detected by using capillary electrophoresis used instead of polyacrylamide gel.

→ Helps in detecting hormone−receptor complexes that bind to DNA at HREs (hormone response elements).

→ Location of TF binding sites in eukaryotic enhancers, operators, and silencers.

→ Detecting the lac repressor responsible for switching off operons in *E. coli*.

3. Gel retardation assay or electrophoretic mobility shift assay or band shift assay

Central to the DNA functions is its interaction with a wide range of proteins, which control various cellular processes such as DNA replication, DNA repair and recombination, transcription, and regulation of all the corresponding processes. Consequently, studying DNA−proteins interaction is critical to explore the dynamics of these phenomena. The Gel retardation assay is one such technique employed to explore gene regulation by studying protein—DNA interactions (Buratowski & Chodosh, 1996; Carey, 1991; Fried & Garner, 1998; Fried, 1989; Garner & Revzin, 1986; Lane et al., 1992). This technique was created by Fried and Crothers (1981) and Garner and Revzin (1981). This technique is versatile in its ability to resolve various conformational and stoichiometric complexities of proteins complexed with specific DNA sequences. Moreover, this technique facilitates the studying of DNA—protein interactions in whole-cell extracts or crude extracts, purified preparations, and transcripts obtained from in vitro protocols. Consequently, overall electrophoretic mobility shift assay (EMSA) facilitates the identification of sequences bound by transcription factors and other regulatory proteins binding at regulatory sites in upstream regions of core as well as proximal promoter regions of genes. Additionally, kinetic and thermodynamic parameters are quantitatively studied in detail. The technique facilitates the detection of TFs in protein fractions incubated with DNA fragments labeled with the radioactive active substance. In addition, EMSA is used to study the interaction of multiple proteins to a DNA molecule. In addition, there are several variants of EMSA to analyze and detect the protein—DNA interactions expanding its dynamic applications (Hellman & Fried, 2007). Moreover, the modified version of EMSA a nonradioactive based technique is utilized to study the RNA—protein interactions (Daras et al., 2019).

3.1 Basic working principle (Fig. 4.2)

The basic principle behind EMSA technique lies in the slow migration of protein—DNA complexes in comparison to the free linear fragments of DNA when electrophoresed in agarose gel or polyacrylamide. During the process, the DNA migration is retarded or shifted. When associated with proteins, the technique is known as a gel shift or gel retardation assay. During the process of electrophoresis, the protein—DNA complexes are quickly resolved from unbound DNA providing the critical snapshot differentiating DNA—protein complex from free DNA fragments. The protein—DNA complexes are stabilized by caging effect of the gel matrix and the low ionic strength of the buffer used in electrophoresis. The length of DNA fragments used in EMSA ranges from 100 to 500 bps usually obtained by digestion of larger DNA molecules by using restriction endonucleases or through PCR amplification. It is important to employ different controls in EMSA. For instance, the running of scrambled DNA—protein to make clear that protein—DNA interaction is specific. Moreover, the monoclonal antibody is added to the labeled DNA—protein complex. EMSA is very important for demonstrating the direct interaction of the protein with specific DNA sequences. The outcome of this technique does not reflect the actual in vivo DNA—protein interaction mechanism. For this purpose, it is critical to further unveil the DNA—protein interactions with other feasible techniques such as ChIP. The general workflow EMSA is as follows (Fig. 4.3).

3.2 **Variants of EMSA**

✔ **Double-label assays and continuous variation**
The variant of EMSA employed to determine the DNA—protein binding stoichiometry

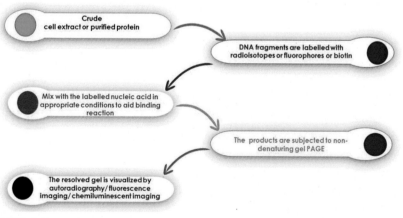

FIGURE 4.3

The general workflow of electrophoretic mobility shift assay.

✔ **Gel Shift Assays (GSA) followed by high-temperature electrophoresis**
Identification of DNA fragments containing specific binding sites for a given protein

✔ **Time course**
It is used to measure the association and/or dissociation kinetics of DNA—protein interactions

✔ **Combined GSA and footprinting**
Reduction of signal from free nucleic acid in footprint pattern

✔ **Circular permutation**
The technique is used to detect the bending of DNA probably due to protein binding

✔ **GSA followed by electron microscopy**
Employed to analyze the conformations formed by protein—DNA complexes

✔ **GSA followed by** Sodium Dodecyl Sulphate (**SDS**)—**polyacrylamide gel electrophoresis with western blot detection/MS**
Aids in identifying the unknown proteins bound to DNA

✔ **Phased bends analysis**
The modification is employed to determine the direction of a protein-induced bend with respect to a standard bend locus

✔ **Binding partition analysis**
Helps to evaluate the binding cooperativity for DNA with multiple binding sites of proteins

✔ **Reverse EMSA**
Aids in DNA binding and measuring the binding affinity

✔ **Cryogenic EMSA**
Helps in detection of protein—DNA complexes that are labile

✔ **Topoisomer EMSA**
The protein—DNA interactions detected supercoiled DNA

✔ **Antibody supershift**
The variant is used to identify proteins that carry a specific epitope in mobility-shifted complexes

✔ **Nucleosome shift**
Helps in the detection of nucleosome-binding protein

3.3 Applications of EMSA

3.3.1 To quantify the TF-binding sites and nuclear receptors in DNA

The technique used to examine TF binding sites as well as nuclear receptors to the specific DNA regions (Soshilov & Denison, 2014).

3.3.2 Estimating degree of biotinylated DNA

A simple nonradioactive EMSA can be used to resolve the weakly biotinylated DNA according to the number of labeled biotin molecules. Hence, EMSA can easily detect the average distribution and number of biotinyl residues in the target photobiotinylated RNA/DNA.

3.3.3 Cancer detection

The technique has immensely contributed to unravel the deregulated molecular and metabolic pathways in a wide range of cancers (Ramteke et al., 2019).

4. Chromatin immunoprecipitation and its variants

The ChIP is another molecular biology technique used to study/investigate the DNA—protein interactions. This technique was created by Gilmour and Lis (1985), when studying the interaction of RNA polymerase II with the promoter sequences in the DNA of *Drosophila melanogaster.* In this technique, the starting material is derived from the cultured cells or fresh tissues directly. Even though it is difficult to exactly determine whether the protein of interest has interacted with DNA directly or is complexed to be part of a bigger complex formed by other proteins. The protein of interest is selectively precipitated by the antibodies from chromatin within the fixed cell to find out the interacting DNA sequence. It is because of this basic principle that ChIP is a widely sued technique for mapping the localization of transcription factors, chromatin-modifying enzymes, histone variants, or posttranslationally modified histones in specific locations of the genome under investigation. Amid all these applications, ChIP analysis remained a cumbersome protocol since the technique needed a huge number of cells. Simultaneously, scientists modified certain working elements of ChIP to overcome these limitations. Moreover, the studies of ChIP are further combined with high throughput sequencing technologies and DNA microarray to further widen the applications (Collas, 2009). For instance, the predictive and distinct chromatin signatures pertaining to enhancers and transcriptional sites in the human genome have been considerably explored by ChIP analysis (Heintzman et al., 2007). Further, the dynamics of the chromatin state with respect to mapping and analysis is greatly explored by ChIP analysis (Ernst et al., 2011).

The technique starts with the cross-linking of DNA and protein by treating living cells with UV light or chemical linkers (Collas, 2009; Cai & Huang, 2012). The step is followed by the extraction of cross-linked chromatin followed by fragmentation either by digestion or sonication to obtain fragmented cross-linked chromatin. Finally, the immunoprecipitated DNA is released from the cross-linked protein, purified, and then subjected to PCR/rtPCR/Southern blotting/sequencing/cloning methods (Fig. 4.4) (Cai & Huang, 2012).

4.1 Workflow of ChIP in detail (Fig. 4.5)

4.1.1 Step 1: protein—DNA cross-linking within cells

In this step, protein-chromatin stabilization is accomplished by covalently cross-linking under in vivo conditions. Chemicals such as formaldehyde in combination with EGS (ethylene glycol bis, succinimidysuccinate) and DSG cross-linker (disuccinimidyl glutarate; di(N-succinimidyl) glutarate) (homo-bifunctional cross-linking agent easily permeable to membranes). The transient protein—DNA interactions are usually transient that is why it is critical to stabilize interaction by cross-linking. Formaldehyde alone cannot act as an efficient cross-linking agent, as it is a zero-length cross-linker (a reagent that couples a molecule with another through covalent bonds directly by chemical activation). The cross-linking agents such as EGS and DGS easily permeate into the cells and enhance the stabilization of complexes by effectively locking and trapping DNA—protein complexes.

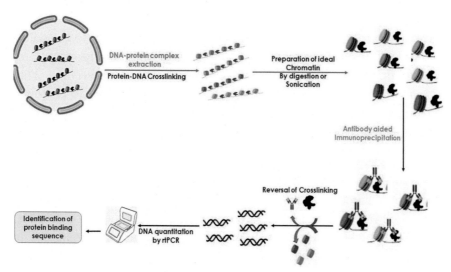

FIGURE 4.4

Chromatin immunoprecipitation—the isolated chromatin with proteins is subjected to fragmentation, followed by Abs against the desired protein. In the next step, the Ab-aided precipitation of desired DNA fragments is obtained which in turn is subjected to reverse cross-linking for removing Ab and proteins to yield naked dsDNA fragments. The DNA fragments are then subjected to qPCR reactions for amplification.

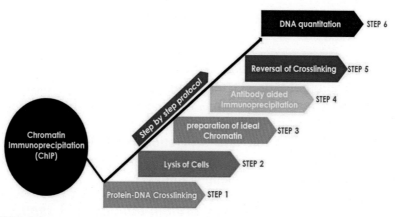

FIGURE 4.5

Outline for steps involved to carry out the ChIP protocol.

4.1.2 Step 2: lysis of cells

This stage brings cross-linked protein—DNA complexes into the solution by cell lysis. During this process, the cellular components are released due to the lysis of the plasma membrane by detergents. The salts and detergents as such aid in the removal of cytosolic proteins and do not affect the protein—DNA complex. Several reagents are used to separate the nuclear fractions from the rest of the cellular components to enhance sensitivity and eliminate the background signal.

4.1.3 Step 3: preparation of ideal chromatin

To clearly analyze the protein binding sites in the genomic DNA, the isolated chromatin must be sheared into smaller fragments of suitable size. Mechanical, enzymatic (micrococcal nuclease), and sonication methods are sued for obtaining an ideal chromatin fragment size of up to 200 to >1000 bps. The micrococcal nuclease-aided digestion is the best suitable method for obtaining highly reproducible fragments.

4.1.4 Step 4: immunoprecipitation of target protein—DNA complex

Immunoprecipitation is one of the critical steps in ChIP. In this step, a monoclonal antibody against the target protein against transcription factor or specific modified histone or some cofactor is designed to aid precipitation. Consequently, this step helps in selecting out specific proteins complexed to the DNA fragment. The antibody—protein—DNA complex is separated from the rest of the mixture by affinity chromatography

4.1.5 Step 5: cross-linking reversal

For quantification of target DNA fragments, the protein—DNA cross-links must be reversed. The step is frequently accomplished by incubations the products at high temperatures or by degrading proteins by proteinase K. Purification and separation of digested proteins is accomplished by phenolo-choloroform treatment step. DNA purification is usually done by using spin columns.

4.1.6 Step 6: quantification of protein binding DNA sequence

The purified DNA is quantified by using quantitative PCR (qPCR).

4.2 The derivatives of ChIP technique

4.2.1 ChIP cloning

In this method, immunoprecipitated DNA from the standard ChIP method is cloned and sequenced. Several modifications were made to adapt the standard ChIP method to clone the immunoprecipitated DNA (Weinmann & Farnham, 2002). The amount of DNA recovered is very less adding to its drawback. The drawback of this technique is overcome by the amplification by PCR. It is important for ChIP cloning to have an appropriate size in chromatin; hence, it is critical to subject chromatin to sonication for obtaining chromatin fragment size of 100−500 bp for precise location of protein binding sites in DNA. DNA sonication results in yielding overhangs at the 5′ and 3′ ends, which are not desirable for ChIP cloning. For rectifying this drawback, T4 DNA polymerase is employed to create the blunt ends, and then these blunt fragments are cloned in vectors.

4.2.2 ChIP-chip/ChIP-on-chip (ChIP-chromatin immunoprecipitation-chip-DNA microarray technology)

This technique is used to isolate and identify all the binding sites of specific DNA binding proteins to the genomes in live cells (Zheng et al., 2007). In this technique after the reversal of cross-linking, all immunoprecipitated fragments of DNA are further amplified by using PCR protocol. The amplified products are labeled with fluorochrome and finally hybridized to a whole genome array. These additional steps in chip−chip leads to the identification of all the sites in the genome where proteins bind. The microarray probes can tile the whole genome to obtain signals of one dimension. The peak shapes represent the protein binding sites. For instance, this technique helps to identify the binding sites of Nanog, Oct4, and Sox2 transcription factors to genomic DNA, hence, identifying distinctive regions in genomic DNA responsible for coding genes pertaining to self-renewal by human embryonic cells (Watson James et al., 2013).

4.2.3 ChIP-CpG microarray

The technique is a combination of ChIP and CpG microarray designed for identifying a large number of protein-binding genomic sites (Iyer et al., 2001; Robyr & Grunstein, 2003). If promoters are to be located, CpG islands are preferable over cDNA microarrays. Most of the eukaryotic promoters consist of CpG islands. The immunoprecipitated DNA from ChIP reactions obtained from the experiments is reverse cross-linked, followed by two-step PCR amplification, viz., in the first reaction using degenerate primer and the second by simple regular PCR reactions. After PCR amplification, the products are labeled with two fluorophores, viz., Cy3 and Cy5. The labeled products are then hybridized to microarray glass slides containing open reading frames or intergenic sequences or both. Then the glass slides are scanned for intensities of fluorescence to determine the pattern of acetylation (Robyr & Grunstein, 2003).

4.2.4 ChIP-Seq

It is the newest version of ChIP with a simple and powerful setup. The DNA fragments are subjected to direct sequencing using next-generation sequencing after DNA bound with proteins is liberated by reverse cross-linking, leading to the detection and measuring of the target DNA sequences. It generates ever-increasing volumes of biological data sets of protein binding sites in DNA. In principle, the techniques allow mapping DNA—protein interactions under in vivo conditions with high resolution and low cost covering whole-genome DNA-binding sites (Furey, 2012; Park, 2009). After following ChIP workflow, the DNA—protein is immunoprecipitated by specific Abs followed by coprecipitation, purification, and high-throughput sequencing. The combinatorial approach of ChIP and next-generation sequencing methods provide great insights into the range of diseases and biological metabolic pathways critical for the development and onset of cancers. The advantage of ChIP-Seq lies in its generation of millions of counts by genome-wide profiling in combination with massively parallel sequencing approaches. In addition, to several advantages mentioned below, ChIP does not require any prior knowledge of genomes.

- ChIP-Seq multiple samples may be processed with cost-effective unbiased and precisely to analyze the epigenetic modifications (Malabarba et al., 2022).
- Defines genome-wide TF-binding sites
- Various DNA samples are compatible for analysis

- Investigates the gene regulatory sites in combination with RNA sequencing and methylation patterns.

4.2.5 ChIP-exo (exo-λ exonuclease)

It is a modified version of the ChIP Seq technique to improve the reduction in background noise as well as a higher resolution during analysis (Rhee & Pugh, 2011). It involves the incorporation of lambda exonuclease digestion in preparation of the library to effectively footprint target DNA sequence. The libraries obtained from ChIP-exo are then subjected to high throughput next-generation sequencing. The techniques are aimed to obtain ultra-high-resolution patterns from a wide range of biological samples such as human cell systems, yeast, mouse, rat, and bacteria (Chen et al., 2014; Cho et al., 2015; Murphy et al., 2015; Wales et al., 2014; Zere et al., 215). This technique is also employed to investigate the organization of the preinitiation complex formed in transcription and to study histones' subnucleosomal structure (Rhee & Pugh, 2012; Rhee et al., 2014). Following are the major applications of ChIP-exo:

- Nucleosomal structure and resolving subnucleosomal structure
- Studying histone H3 modification marks
- Studying asymmetric localization of histone variant H2A.Z
- Studying subnucleosomal asymmetry of histone H2Bub and its machinery
- To investigate the precise organization of chromatin regulatory proteins
- Organization of transcriptional preinitiation complex
- Mapping DNA-binding protein complexes
- High-resolution mapping of protein—DNA interactions
- Identification of protein—DNA interactions in nucleosome-free regions
- Organization of transcriptional preinitiation complex interplaying with chromatin.

4.2.6 RedChIP

The technique combines ChIP and RNA—DNA proximity ligation methods to identify the RNA—chromatin interactions facilitated by proteins. For instance, by using Abs EZH2 subunit of the polycomb repressive complex 2 and CTCF (architectural protein) involved in the repression of genes. Hence, RedChIP is a versatile tool for investigating nuclear ncRNAs (Gavrilov et al., 2022; Patton et al., 2020). RedChIP is versatile in identifying the noncoding RNAs bound to nonhistone proteins or any other protein locations in chromosomes (Gavrilov et al., 2022).

4.2.7 ChIA-PET

This technique is a genome-wide high throughput technique de novo method, which combines ChIP and chromatin conformation capture (3C) technology (Fullwood & Ruan, 2009a). It is an emerging technique to study chromatin interactions with higher resolution associated with proteins for studying functional and 3D chromatin structures (Fullwood et al., 2009a). Further, this technique is utilized to characterize the 3D stricture of chromatin in the nucleus for the identification of transcriptional binding sites in the whole genome (Fullwood et al., 2009a, 2009b, 2009c).

4.3 The ChIP databases

4.3.1 proChIPdb

An interactive database is developed by using ChIP-obtained data. It is the collection of public ChIP-seq/-exo data pertaining to prokaryotes (Decker et al., 2022). The data sets are presented in dashboards that include nucleotide-resolution genome viewers, summary plots, and curated binding sites of prokaryotes. Through this database, users can download all data and follow external links, for understanding regulons through various biological databases and online literature. In addition, users can find target genes where TFs bind and regulate gene expression.

4.3.2 ChIP Atlas

It is a comprehensive and integrative database used to explore the epigenetic data sets developed by DNase-Seq, ChIP-Seq, Bisulfite-Seq data, and ATAC-Seq. In addition, the ChIP Atlas covers all data archived in biological databases such as NCBI, DDBJ, and EBI (Oki et al., 2018).

4.3.3 PRODORIC2 database

It is one of the largest collections of datasets pertaining to DNA binding sites for TFs in prokaryotes (Eckweiler et al., 2018). It is the modified version of the PRODORIC database with more intuitive and higher user-friendly access. Apart from the technical advancements, this database provides more than 1000 binding sites of TFs and additionally 110 weight matrices position sites genome-wide pattern searches. It also includes DNA binding sites generated from high throughput experimentations. Users can access the PRODORIC2 at http://www.prodoric2.de, website.

4.3.4 Encyclopedia of DNA elements (ENCODE) project

This project resulted in systematically mapping the regions of DNA where transcription factor binding sites, chromatin structure, and histone modification (The ENCODE, 2004, 2011). The datasets help us to assign functions for 80% of genomic sites. Furthermore, we can easily assess the mechanism of gene regulations based on existing datasets. In general, the project provides deep insights into the organization and regulation of genomes and vast resources on functional annotations for applied medical research.

4.4 Applications of ChIP

4.4.1 Identifying the binding sites of RNA polymerase II and transcription factors

- By using an antibody specific to RNA polymerase II, ChIP provides additional proof that the transcription occurs at divergent CpG island promoters in the case of mammals. Moreover, ChIP assays were applied to validate the TATA-binding protein (TBP) binds to the CpG islands in the promoter regions of eukaryotes by using labeled antibodies specific to TBP.
  The ChIP in combination with DNA sequencing is used for identifying the binding sites of TFs across the whole genome to understand the regulation of gene expression (Decker et al., 2022).

4.4.2 Identifying the repressed regions of chromosomes

- By using antibodies specific to the acetylated lysines in histone N-terminal tails, one can identify the existence of repressed histones in telomeres and locations of hypoacetylated loci in chromosomes.

4.4.3 Identification of paused transcription by RNA polymerase II

- The paused elongation mediated by RNA polymerase II at 100 bp away from the start site in metazoans is identified by the application of ChIP assays.

4.4.4 Mapping the binding of transcription factors

- ChIP assay helps in mapping the binding sites for TFs and results showed that TF binding site is very coincident with the DNase I hypersensitive sites, the short regions up to 100 bps found in the genomic DNA that are extremely sensitive to the digestion by DNaseI.

4.4.5 Identifying the frequent transcription sites in eukaryotic genomes

- By applying ChIP using antibodies specific to RNA polymerase II, the amount of Pol II in different positions of transcription units can be assessed. This finding will also pave in indicating that Pol II elongates the mRNA more rapidly at certain sites than others (Arnold et al., 2016).

4.4.6 Identifying the repressed transcription sites in eukaryotic genomes

- By using antibodies specific to the acetylated lysines in histone tails of chromatin, ChiP analysis confirms that repressed regions exist in telomeres (Arnold et al., 2016).

4.4.7 Studying epigenetic modifications

The ChIP and its derivatives are widely sued for studying epigenetic modifications in eukaryotic genomes

5. Conclusion

Studying the DNA—protein interactions is important for investigating the cellular and molecular mechanisms such as DNA modifications, transcription and its regulations, and other processes kinked to gene expression. For several decades, scientists are investigating the mechanism by which proteins interact with specific sequences of DNA molecules. A wide range of techniques have evolved to describe the DNA—protein interactions. For instance, EMSA is in vitro technique used to identify the protein—DNA interactions to have insights into the functional studies of genomes for example that of TFs, enhancer proteins, etc. In combination with SDS-PAGE, the techniques can aid in the identification of DNA sequences binding to proteins in crude extracts. For the identification of proteins to the sequence in DNA, DNA footprinting is preferable because this technique is easy and faster. This technique operates both in vivo as well as in vitro conditions. In addition, ChIP assay and its variants are popular tools to investigate the chromatin structures and DNA—proteins critical for transcription. The ChIP assay is important for the identification of DNA-binding proteins critical for the regulation of gene expression. Once performed under appropriate controls, the ChIP is a very versatile technique to explore DNA—protein interactions under in vivo conditions.

References

Arnold, B., Chris, A. K., Harvey, L., Angelika, A., Hidde, P., Anthony, B., Monty, K., & Kelsey, C. M. (2016). *Molecular cell biology* (8th ed., p. 1280). WH Freeman. ISBN-10: 1464183392 ISBN-13: 978-1464183393.

Buratowski, S., & Chodosh, L. A. (1996). In F. M. Ausubel, R. Brent, R. E. Kingston, D. D. Moore, J. G. Seidman, J. A. Smith, & K. Struhl (Eds.), *Current protocols in molecular biology*. New York: John Wiley and Sons. pp. 12.2.11—12.2.10.

Cai, Y. H., & Huang, H. (2012). Advances in the study of protein—DNA interaction. *Amino Acids, 43*(3), 1141—1146.

Carey, J. (1991). Gel retardation. *Methods in Enzymology, 208*, 103—117.

Chen, J., et al. (2014). Single-molecule dynamics of enhanceosome assembly in embryonic stem cells. *Cell, 156*, 1274—1285.

Cho, S., et al. (2015). The architecture of ArgR-DNA complexes at the genome-scale in *Escherichia coli*. *Nucleic Acids Research, 43*, 3079−3088.

Collas, P. (2009). The state-of-the-art of chromatin immunoprecipitation. In P. Collas (Ed.), *Methods in molecular biology (methods and protocols): Vol 567. Chromatin immunoprecipitation assays*. Totowa, NJ: Humana Press. https://doi.org/10.1007/978-1-60327-414-2_1

Daras, G., Alatzas, A., Tsitsekian, D., Templalexis, D., Rigas, S., & Hatzopoulos, P. (2019). Detection of RNA-protein interactions using a highly sensitive non-radioactive electrophoretic mobility shift assay. *Electrophoresis, 40*(9), 1365−1371.

Decker, K. T., Gao, Y., Rychel, K., Al Bulushi, T., Chauhan, S. M., Kim, D., & Palsson, B. O. (2022). proChIPdb: A chromatin immunoprecipitation database for prokaryotic organisms. *Nucleic Acids Research, 50*(D1), D1077−D1084.

Eckweiler, D., Dudek, C. A., Hartlich, J., Brötje, D., & Jahn, D. (January 4, 2018). PRODORIC2: The bacterial gene regulation database in 2018. *Nucleic Acids Res, 46*(D1), D320−D326. https://doi.org/10.1093/nar/gkx1091. PMID: 29136200; PMCID: PMC5753277.

The ENCODE (ENCyclopedia of DNA elements) project. ENCODE project consortium. *Science, 306*(5696), (October 22, 2004), 636−640.

Ernst, J., et al. (2011). Mapping and analysis of chromatin state dynamics in nine human cell types. *Nature, 473*, 43−49.

Fried, M. G. (1989). Measurement of protein-DNA interaction parameters by electrophoresis mobility shift assay. *Electrophoresis, 10*, 366−376.

Fried, M., & Crothers, D. M. (1981). Equilibria and kinetics of lac repressor-operator interactions by polyacrylamide gel electrophoresis. *Nucleic Acids Research, 9*(23), 6505−6525.

Fried, M. G., & Garner, M. M. (1998). In D. Tietz (Ed.), *Molecular biology methods and applications* (pp. 239−271). Elsevier.

Fullwood, M. J., et al. (November 5, 2009). An oestrogen-receptor-alpha-bound human chromatin interactome. *Nature, 462*(7269), 58−64. https://doi.org/10.1038/nature08497. PMID: 19890323; PMCID: PMC2774924.

Fullwood, M. J., & Ruan, Y. (2009a). ChIP-based methods for the identification of long-range chromatin interactions. *Journal of Cellular Biochemistry, 107*, 30−39. https://doi.org/10.1002/jcb.22116

Fullwood, M. J., Wei, C. L., Liu, E. T., & Ruan, Y. (2009b). Next-generation DNA sequencing of paired-end tags (PET) for transcriptome and genome analyses. *Genome Research, 19,* 521–532. https://doi.org/10.1101/gr.074906.107

Furey, T. S. (2012). ChIP-seq and beyond: New and improved methodologies to detect and characterize protein-DNA interactions. *Nature Reviews Genetics, 13*(12), 840–852.

Galas, D. J., & Schmitz, A. (1978). DNAse footprinting: A simple method for the detection of protein-DNA binding specificity. *Nucleic Acids Research, 5,* 3157–3170.

Garner, M. M., & Revzin, A. (1981). A gel electrophoresis method for quantifying the binding of proteins to specific DNA regions: Application to components of the *Escherichia coli* lactose operon regulatory system. *Nucleic Acids Research, 9*(13), 3047–3060.

Garner, M. M., & Revzin, A. (1986). The use of gel electrophoresis to detect and study nucleic acid-protein interactions. *Trends in Cell Biology, 11,* 395–396.

Gavrilov, A. A., Sultanov, R. I., Magnitov, M. D., Galitsyna, A. A., Dashinimaev, E. B., Aiden, E. L., & Razin, S. V. (2022). RedChIP identifies noncoding RNAs associated with genomic sites occupied by Polycomb and CTCF proteins. *Proceedings of the National Academy of Sciences, 119*(1).

Gilmour, D. S., & Lis, J. T. (1985). In vivo interactions of RNA polymerase II with genes of *Drosophila melanogaster. Molecular and Cellular Biology, 5*(8), 2009–2018.

Heintzman, N. D., et al. (2007). Distinct and predictive chromatin signatures of transcriptional promoters and enhancers in the human genome. *Nature Genetics, 39,* 311–318.

Hellman, L., & Fried, M. (2007). Electrophoretic mobility shift assay (EMSA) for detecting protein–nucleic acid interactions. *Nature Protocols, 2,* 1849–1861. https://doi.org/10.1038/nprot.2007.249

Iyer, V. R., Horak, C. E., Scafe, C. S., Botstein, D., Snyder, M., & Brown, P. O. (2001). Genomic binding sites of the yeast cell-cycle transcription factors SBF and MBF. *Nature, 409,* 533–538.

Lane, D., et al. (1992). Use of gel retardation to analyze protein-nucleic acid interactions. *Microbiological Reviews, 56,* 509–528.

Malabarba, J., Chen, Z., Windels, D., & Verdier, J. (2022). Chromatin Immunoprecipitation dataset of H3ac and H3K27me3 histone marks followed by DNA sequencing of *Medicago truncatula* embryos during control and heat stress conditions to decipher epigenetic regulation of desiccation tolerance acquisition. *Data in Brief,* 107793.

Murphy, M. W., et al. (2015). An ancient protein-DNA interaction underlying metazoan sex determination. *Nature Structural & Molecular Biology, 22*, 442−451.

Ofran, Y., Mysore, V., & Rost, B. (2007). Prediction of DNA-binding residues from sequence. *Bioinformatics, 23*(13), i347−i353.

Oki, Shinya, Ohta, Tazro, et al. (2018). ChIP-Atlas: A data-mining suite powered by full integration of public ChIP-seq data. *EMBO Reports*, e46255. https://doi.org/10.15252/embr.201846255

Park, P. J. (2009). ChIP-seq: Advantages and challenges of a maturing technology. *Nature Reviews Genetics, 10*(10), 669−680.

Patton, S. R., Clements, M. A., Marker, A. M., & Nelson, E. L. (2020). Intervention to reduce hypoglycemia fear in parents of young kids using video-based telehealth (REDCHiP). *Pediatric Diabetes, 21*(1), 112−119.

Ramteke, P., Athavale, D., & Bhat, M. K. (2019). Dysregulated signaling pathways in cancer: Approaches and applications. In K. Bose, & P. Chaudhari (Eds.), *Unravelling cancer signaling pathways: A multidisciplinary approach*. Springer. https://doi.org/10.1007/978-981-32-9816-3_10

Ren, B., Robert, F., Wyrick, J. J., Aparicio, O., Jennings, E. G., Simon, I., Zeitlinger, J., Schreiber, J., et al. (2000). Genome-wide location and function of DNA binding proteins. *Science, 290*, 2306−2309.

Rhee, H. S., Bataille, A. R., Zhang, L., & Pugh, B. F. (2014). Subnucleosomal structures and nucleosome asymmetry across a genome. *Cell*, 1377−1388.

Rhee, H. S., & Pugh, B. F. (2011). Comprehensive genome-wide protein-DNA interactions detected at single-nucleotide resolution. *Cell, 147*, 1408−1419.

Rhee, H. S., & Pugh, B. F. (2012). Genome-wide structure and organization of eukaryotic pre-initiation complexes. *Nature, 483*, 295−301.

Robyr, D., & Grunstein, M. (2003). Genome-wide histone acetylation microarrays. *Methods, 31*, 83−89.

Soshilov, A. A., & Denison, M. S. (2014). DNA binding (gel retardation assay) analysis for identification of aryl hydrocarbon (Ah) receptor agonists and antagonists. In G. Caldwell, & Z. Yan (Eds.), *Optimization in drug discovery. Methods in pharmacology and toxicology*. Humana Press. https://doi.org/10.1007/978-1-62703-742-6_12

Spencer, V. A., Sun, J. M., Li, L., & Davie, J. R. (2003). Chromatin immunoprecipitation: A tool for studying histone acetylation and transcription factor binding. *Methods, 31*, 67−75.

A user's guide to the encyclopedia of DNA elements (ENCODE). ENCODE project consortium. *PLoS Biol, 9*(4), (April 2011), e1001046.

Wales, S., Hashemi, S., Blais, A., & McDermott, J. C. (2014). Global MEF2 target gene analysis in cardiac and skeletal muscle reveals novel regulation of DUSP6 by p38MAPK-MEF2 signaling. *Nucleic Acids Research, 42*, 11349–11362.

Watson James, D., Baker Tania, A., Bell Stephen, P., Alexander, Gann, Levine, Michael, & Richard, Losick (2013). *Molecular biology of the gene (2-downloads)* (7th ed.). Kindle Edition.

Weinmann, A. S., Bartley, S. M., Zhang, T., Zhang, M. Q., & Farnham, P. J. (2001). Use of chromatin immunoprecipitation to clone novel E2F target promoters. *Molecular and Cellular Biology, 21*, 6820–6832.

Weinmann, A. S., & Farnham, P. J. (2002). Identification of unknown target genes of human transcription factors using chromatin immunoprecipitation. *Methods, 26*, 37–47.

Weinmann, A. S., Yan, P. S., Oberley, M. J., Huang, T. H., & Farnham, P. J. (2002). Isolating human transcription factor targets by coupling chromatin immunoprecipitation and CpG island microarray analysis. *Genes Dev, 16*, 235–244.

Yan, Y., Kluz, T., Zhang, P., Chen, H. B., & Costa, M. (2003). Analysis of specific lysine histone H3 and H4 acetylation and methylation status in clones of cells with a gene silenced by nickel exposure. *Toxicology and Applied Pharmacology, 190*, 272–277.

Zere, T. R., et al. (2015). Genomic targets and features of BarA-UvrY (-SirA) signal transduction systems. *PLoS One, 10*, e0145035.

Zheng, M., Barrera, L. O., Ren, B., & Wu, Y. N. (2007). ChIP-chip: Data, model, and analysis. *Biometrics, 63*(3), 787–796. https://doi.org/10.1111/j.1541-0420.2007.00768.x. PMID: 17825010.

Nutrigenomics: Insights into the influence of nutrients on functional dynamics of genomes

1. Introduction

The physiological, metabolic, and molecular dynamics in animals is amended by organic and inorganic nutrients. All the critical process happening in the cells is controlled by the working manual hidden in the genome, which drives gene expression controlled by biomolecules such as proteins. The field of science that deals with investigating the role of nutrients on expression levels of genomes is known as nutrigenomics. Nutrigenomics puts a special emphasis on the interaction of bioactive compounds on the functional dynamics of the genome (Farhud et al., 2010). Nutrigenomics is employed to study the effect of nutrients on genomes to cover nutrient-based diseases. On a global scale, millions of humans are suffering from malnutrition or deficiency linked to a daily intake of food. For instance, a large number of deaths at the global level are caused by diseases such as diabetes, hypertension, blood pressure, cancers, and cardiovascular diseases that are intimately linked to nutrients. Remarkably, nutrigenomic studies help to understand the diet-related disruptions in signaling cascades governing the diseases such as diabetes, cancer, hypertension, and cardiovascular disorders. For instance, Rickets and osteomalacia are the outcomes of foods lacking a considerable amount of vitamin D. On the other hand, a higher intake of vitamin D is found associated with the development of cancers because vitamin D is critically involved in the regulation of growth and proliferation of cells. In addition, vitamin D is found to regulate its activity through interaction with the vitamin D receptor (VDR), a class of nuclear hormone receptors to control cellular proliferation and division. The studies also revealed that VDR polymorphism is regulated by the dietary outcome and plays a critical role in the development of diverse diseases in humans. Another example is to understand the role of nutrigenomics to study the impact of dietary fibers on the polymorphism in the angiotensinogen gene that controls blood pressure in humans. Furthermore, studies also revealed that minerals play a pivotal role in shielding humans from cancers, for example, the effect of selenium in glutathione peroxidase production through its action on membrane integrity, hydrogen peroxide (Almendro & Gascan, 2012). In addition, the nutrigenomic studies also revealed the role of zinc in regulating gene expression, genomic stability, and modulation of apoptosis (Almendro & Gascan, 2012). In recent decades, scientists have shifted their focus on utilizing orphan or underutilized crops such as amaranth (*Amaranthus* spp.), buckwheat (*Fagopyrum esculentum* Moench.), Chia (*Salvia hispanica* L.), and Quinoa (*Chenopodium quinoa* Wild) due to their higher nutritional qualities. These crops are highly rich in minerals, amino acids, vitamins, bioflavonoids, and phenolics (Bekkering & Tian, 2019; Gebremariam et al., 2014; Singh & Sharma, 2017) and have been used to circumvent a diverse range of diseases (Fig. 5.1). For instance, amaranth and quinoa grains contain an optimal composition of proteins almost similar to the protein and essential amino acid composition of cow milk. Moreover, bioactive compounds are also identified in amaranth and chia (Grancieri et al., 2019; Silva-Sánchez et al., 2008). Remarkable to underutilized crops is the absence of gluten in grains, which is medically suitable for treating patients having celiac disease. Consequently, reports regarding nutrigenomics clearly indicate that studies are aimed at investigating the interactions of nutrients with genes and then exploring the precise source of nutrient-rich food crops for curing a diverse range of dietary diseases in humans (Corella & Ordovas, 2009; Ferguson, 2009; Simopoulos, 2010) (Fig. 5.2). Keeping in view the importance of nutrigenomics and its investigation of overcoming health complications by utilizing functional foods, we aimed to include this chapter in our book to have great insights and applicability of genomics in disease biology.

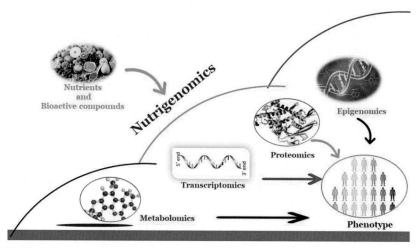

FIGURE 5.1

Technical approaches of nutrigenomics to study the relationship of nutrients and their impact on gene expression.

Scope of plant-based foods in addressing the "life-style" diseases

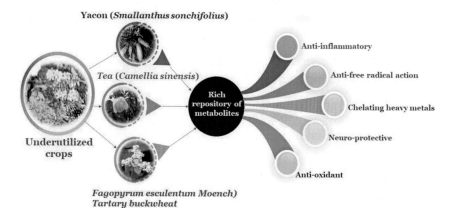

FIGURE 5.2

Capacity of underutilized crops as a frontline repository of bioactive compounds such as amino acids, proteins, dietary fibers, antioxidant vitamins, omega-3-fatty acids, prebiotics, and flavonoids to treat diseases like hypertension, angina, arrhythmias, hyperlipidemias, and congestive heart failure.

1.1 The beginning of nutrigenomics

The influence of nutrients and diet on health is evolved since ancient times. The field is evolved to study the interaction of foods with genes, in particular inherited genes. For instance, it is recommended that humans suffering from phenylketonuria, a metabolic disorder, are recommended to avoid foods that are rich in phenylalanine amino acid. Similarly in lactose intolerance, a condition in which the adults are not tolerant to catabolize lactose, the patients are advised to intake major milk products. In this hereditary disorder, the gene encoding the lactase enzymes is mutated or turned off after weaning. The field of nutrigenomics jump started after the completion of human genome project around 2001. By 2007, several scientists started unraveling the interrelationship of nutrition, genes, and disease to diversify the field of medical science. Consequently, the introduction of nutrigenomics opened new avenues in nutritional science with the advent of new terminologies and techniques with side-lined combined high throughput technical approaches to study the impact of foods on gene expression. The field is actually a collaborative effort to include diverse scientific fraternities like geneticists, public health officials, and food science and culinary officials. In the current scenario of growing food-based diseases, nutrigenomics may prove a dcfcnse and ultimate solution on a global scale. In the current scientific world, nutrigenomics is carried out in different fields such as nursing, medicine, molecular biology, biochemistry, genetics, and biological and agricultural sciences. Consequently, the field of nutrigenomics evolved to characterize the gene products and their physiological functions and interactions by using genomes, transcriptomes, proteomes, metabolomes, and other omics information to unravel the link between nutrients and genes to get insights into regimes on human health conditions.

1.2 Nutrigenomics research tools

Nutrigenomics is highly explored in the postgenomic era and has opened a new window for nutrition biologists to screen the genetic background of patients influenced by their diet. For instance, microarray technologies mainly including transcriptomics have led to unraveling the physiological effects of dietary proteins, lipids such as omega-3 polyunsaturated fatty acids and dietary conditions of colon cancer patients. In addition, DNA microarray and polymerase chain reaction (PCR) leads to evaluating the interaction of genes and diet to assess gene expression. Additionally, proteomic techniques such as 2D (two-dimensional electrophoresis) were used to evaluate the protein content of poultry meat and egg, assessment of dietary methionine, the toxicity of dioxin, and also assessment regarding the safety of transgenic crops for checking the suitability as a dietary protein source. In addition, metabolomic analysis led to the detection of any changes in the biochemical profile of urine and plasma in rats and other animals for assessing the effect of dirty nutrients. Microarray technologies have been employed to improve the quality of food and its safety in the meat as well as the dairy industry. The application of DNA chip technology has helped in the screening of a large number of genes providing comprehensive knowledge of genetic variants and their regulatory networks influenced by dietary nutrients.

1.3 Markers for nutrigenomics: the assessment of single nucleotide polymorphism

The influence of dietary nutrients on the transcriptomic, proteomic and metabolomic profile was assessed by exploring the variation in genetic polymorphism single nucleotide polymorphism (SNPs). In addition, the gene expression influenced by nutrients was greatly studied to have insights into the diet on human health. A large number of relatively common SNPs have been identified by genomic techniques largely derived from dietary nutrients. For instance, the SNPs were identified as responsible for the development of organ dysfunctioning in humans largely influenced by low choline diets in humans. Likewise, very common SNPs were found in premenopausal women upon dietary intake of low choline diets. Low choline diets have been found to effect fetal brain development in animals such as experimented in rodent models. A large number of SNPs have been identified to assess the diversity linked to diseases such as Down syndrome, cancer, cardiovascular diseases, neural tube defect, leukemia, spina bifida, obesity, and diabetes. Moreover, the SNPs linked to folate metabolic pathways in healthy populations in

India have been explored at a larger scale. PCR-Simple sequence repeats (SSR) microsatellite markers have been used to gain better insights into genotypes of silkworm breeds/hybrids influenced by dietary nutrients.

1.4 Human nutrition and diseases

Once the nutrients of the human body are not supplied properly the functions of cells, tissues and organisms are impaired resulting in several ailments. Consequently, an unbalanced diet causes certain diseases and illnesses and metabolic disorders. For instance, over-eating of foods rich in fats and poor in essential nutrients leads to the development of obesity and the emergence of irritable bowel syndrome (IBS) and gastroesophageal reflux disease (GERD). In addition, diseases such as celiac disease inhibit the human body from getting nutrients from the ingested food. Other diseases linked to the intake of nutrient-poor unbalanced foods include cancer, stomach ulcers, Crohn's disease, kidney diseases, liver diseases, food allergies, blood pressure, and their linked ailments in human organ systems. We have provided a brief account of some nutrient-related diseases for highlighting the importance of nutrients in our daily food intake.

1.5 Celiac disease

Celiac disease is an autoimmune disorder caused by the untoward immune reaction by biomolecules such as gluten protein in the small intestine of humans. The gluten is formed from gliadin and glutenin in presence of water. Both these proteins are used in making bread from rye, wheat, and barley. Mostly commonly celiac disease is found in patients with autoimmune thyroid disease, Down syndrome and Turner syndromes, and type 1 diabetes. The major symptoms of celiac disease include paleness, abdominal pain, gastrointestinal upset, weight loss, loose stool, and poor development in children. Due to the destruction of absorptive surfaces in the small intestine, very less nutrients are absorbed resulting in malnutrition and the evolution of several diseases linked to nutrition. For instance, less absorption of iron and folic acid leads to anemia. Moreover, vitamin D and calcium poor absorption result in osteoporosis characterized by brittle bones. Effective treatments include the consumption of gluten-free foods. So consuming foods based on plants with poor proteins such as corn, millet, buckwheat, and oats are viable options.

1.6 Irritable bowel syndrome

Irritable bowel syndrome is characterized by constipation, bloating, diarrhea, and abdominal pain due to muscle spasms in the colon. IBS is largely caused by mental as well as physical stress and an unbalanced diet. The unbalanced food intake such as that of added sugars, foods high in oils and fats, chocolate, and peppermint intensifies the symptoms of IBS. Similar to GERD, the IBS clinical futures further worsen upon intake of beverages containing alcohol or caffeine. The treatment is based on adjustments in diet and lifestyle modifications. Further recommendations include drinking more water, regular exercise, fiber-rich foods, and a slower rate of eating. In addition, drugs against depression and relaxing the colon may relieve the pain caused by IBS.

1.7 Gastroesophageal reflux disease

GERD is the condition in which the stomach leaks backward the acidic contents into the esophagus and leads to irritation and is usually termed as acid refluxes, which usually occurs once a week. The disease is most frequently found in patients with obese conditions. The symptoms of GERD include backflow of acidic contents in the mouth also called as regurgitation, heart burns, trouble in swallowing food, and frequent occurrence of coughing. Consequently, GERD patients are sensitive to certain specific foods such as fried foods, chocolate, spicy foods, garlic, and tomato. In addition, drinks containing caffeine or alcohol further worsen the symptoms of GERD. As results patient with GERD have inflamed esophageal tissues. The treatment approaches are largely based on dietary adjustments and additionally lifestyle modifications. Moreover, GERD patients are suggested to lose weight, eat light foods, etc. The specific nutritional deficiencies in GERD are largely due to zinc and magnesium. Nowadays, functional foods are emerging options to cure GERD.

1.8 **Food allergies**

It is critical to pay attention to the diversity of foods with respect to their acceptance by individual humans, as every person responds to different foods independent of origin or hereditary disposal. Consequently, we have noticed that several humans are allergic to some chosen foods. A person allergic to certain foods such as peanuts evoke an immune response to these allergic foods through the generation of IgE antibody and subsequent type I hypersensitivity reactions that may be life threatening. The symptoms of food allergy develop very quickly and include Stomach cramps, tingling mouth, difficulty breathing, swelling tongue and/or throat, diarrhea, hives, vomiting, drop in blood pressure, loss of consciousness, and in some cases, a person may die due to severe allergy. The only viable treatment for food allergy is avoiding specific foods, which may be controlled by treatment with epinephrine to regulate the severe allergic reactions. The major foods causing allergies include milk, peanuts, soy, eggs, fish shellfish, tree nuts, and wheat. A comprehensive guideline has been issued by the National Institute of Allergy and Infectious Diseases in collaboration with international agencies regarding the existing list of foods allergic to a certain class of humans.

1.9 **Oral disease**

Oral health is the most critical health factor that may lead to several ailments largely due to poor oral hygiene. It grossly refers to the health of gums, teeth, and associated tissues such as nerves, chewing muscles, ligaments, and salivary glands. Foods such as desserts, fruit juices, soft drinks, and candies are responsible for poor oral health. These foods result in the best substratum for several bacterial species which leads to dental decay. The metabolized substrate by bacterial species results in the production of acids, which in turn damages the teeth enamel and other bone tissues. Further, the acids create acid holes and also damage the gums. Extreme conditions will finally lead to gingivitis.

1.10 **Lifestyle-associated diseases**

The diversified lifestyle and unhealthy habits lead to higher mortality rates on a global scale. The food-based diseases are due to disturbed environments, unhealthy diets, and sedentary lifestyles. Further, on a global scale, about 60% of estimated deaths occur due to chronic diseases linked either the improper food intake or an ill lifestyle (WHO, 2008). The common risk factors include nontransmissible, slow progression, and noninfectious (WHO, 2011). Most of these ailments are linked to metabolic imbalances such as inflammation like epidemiological characteristics, oxidative stress, and mitochondrial alterations. Moreover, the metabolic networks of blood are linked to poor nutrition and an unhealthy diet (Khatibzadeh et al., 2012). It is reported that inflammation is upregulated by a wide range of environmental stressors and unusual dietary schedules. Consequently, the class of foods consumed is responsible for the health and diseased state of humans. The right foods helping to circumvent the symptoms of the disease are generally called as functional foods. For instance, several crop plants are a source of important biomolecules such as amino acids, proteins, dietary fibers, antioxidant vitamins, omega-3-fatty acids, prebiotics, and flavonoids to treat several diseases like hypertension, angina, arrhythmias, hyperlipidemias, and congestive heart failure (Ramaa et al., 2006) (Fig. 5.2). For instance, curcumin a dietary flavonoid acts as an antiadipogenic role in murine and human adipocytes (Kim et al., 2011). Moreover, the induction of IL-6 secretion by leptin and lipopolysaccharide was found to be activated by curcumin. Curcumin also induces a reduction in Tumor necrosis factor (TNF) levels, insulin sensitivity in T2DM-male Sprague Dawley rats, and induction of hypoglycemia (El-Moselhy et al., 2011). The hypoglycemic effects were also induced by steroids such as 4- hydroxyisoleucine (amino acid) and saponintrigonelline extracted from *Trigonella foenumgraecum* (fenugreek). Likewise, omega-3 fatty acids help to regulate cancers and cardiovascular diseases in humans (Laviano et al., 2013; Vrablík et al., 2009). In addition, the linolenic acid found in grilled meat is found to possess anticancer effects (Kelley et al., 2007).

2. Nutrigenomics studies of selected crop plants—a repository of functional foods

Current agricultural system witnesses the shift of crop production from mainstream crops to allied crops such as underutilized or

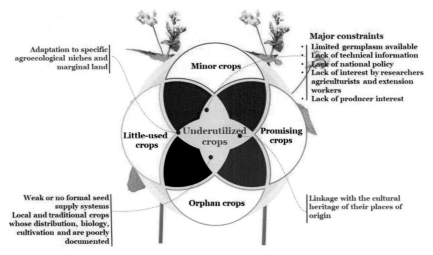

FIGURE 5.3

Pictorial demonstration of limitations and hindrances associated with the cultivation yield and production of underutilized crops for food and nutritional security.

orphan crops. These crops possess several properties of functional foods but are still cultivated on large scale due to limitations in yield, production, and processing (Fig. 5.3). To augment the process of cultivation, the last decade witnessed speed in crop genomics through the intervention of high throughput technologies such as next generation sequencing (NGS) (Varshney et al., 2009). Consequently, NGS has been used to sequence the genes related to nutrients and their link with diseases. Further whole genome sequencing data repository can be used to enhance the understanding of genomic predictions linked to the nutrients (Hickey et al., 2017). These techniques led to explore the genomic information of several underutilized crops such as amaranth, quinoa, buckwheat, broomcorn millet, and teff. The genomics information of these crops unveils the information pertaining to genomic structural variants and SNPs. With this information, scientists can edit desired genes associated with underutilized crops. For instance, certain nutrient-related genes from underutilized crops have been transformed into mainstream crops.

2.1 **Buckwheat**

In amaranth, the seed albumin gene (*AmA1*) encoding the nonaller-
genic protein, whcih is rich in essential amino acids can greatly help
to overcome nutritional requirements in humans. It is reported that
the transfer of this gene in potatoes enhanced the amino acid content.
Likewise, *waxy gene* or granule-bound starch synthase (*GBSS*) was
also reported in amaranth grains. The *PAL* gene (Phenylalanine
Ammonia Lyase) identified in buckwheat (*Fagopyrum* sp.) is medici-
nally important since the enzyme encoded by this gene is critical for
the biosynthesis of two important bioactive compounds such as rutin
and quercetin. Both of these compounds are reported to possess anticancer and antidiabetic properties (Thiyagarajan et al., 2016). Accordingly, the medicinal values of buckwheat can be improved by genetic
engineering techniques (Thiyagarajan et al., 2016). Other important
genes reported in buckwheat include those encoding for flavonoid
biosynthesis, *GBSS* gene, and 2S albumin (Yasui et al., 2004).

2.2 **Quinoa (*Chenopodium quinoa* wild)**

Quinoa is a herbaceous plant belonging to the amaranth family and is
cultivated for its edible seeds. The seeds are rich in minerals, vitamin
Bs, proteins, and dietary fiber. It is important for pseudocereals having
medicinal and nutraceutical properties. For instance, the genes related
to nutritive proteins include *GBSS1A* and *GBSS1B,* genes related to
saponin biosynthesis (Balzotti et al., 2008; Brown et al., 2014). In
addition, genes such as Amaranthin, biosynthesis of betanin, celosia-
nin II, and betalain have been abundantly found in Quinoa (Imamura
et al., 2019). A large number of genetic annotations to collect large-
scale Expressed sequence tags (EST) are reported in seed tissues of
Quinoa linked to express the saponins. Moreover, microarray analysis
led to the identification of several genes linked to saponin biosynthe-
ses such as those coding for cytochrome P450s, cytochrome P450
monooxygenases, and glycosyltransferases. In addition, protein accumulation in quinoa was detected by using techniques such as SDS-
PAGE analyses and real-time and reverse transcriptase PCR (Balzotti
et al., 2008). Consequently, these reports suggest that quinoa is a good
source of nutrition and needs to be further explored through nutrigenomics approaches.

2.3 Amaranth (*Amaranthus* spp.)

Amaranth is the collective term given to plant species belonging to the genus *Amaranthus*, which are short-lived perennial or annual plants. The genus comprises about 400 species found in diverse habitats in temperate, subtropical, and tropical climate zone globally. The species cultivated for grain production include *Amaranthus caudatus, Amaranthus cruentus,* and *Amaranthus hypochondriacus*. These Amaranth species are referred to as "pseudo-cereals" and third-millennium crop plants. Other species such as *Amaranthus blitum, Amaranthus dubius, Amaranthus spinosus, Amaranthus tricolor,* and *Amaranthus viridis* are mostly cultivated for obtaining leafy vegetables. On the other hand, species such as *A. viridis*, *A. dubius*, *A. retroflexus*, *A. spinosus*, *A. hybridus,* and *A. graecizans* exist as weeds. Consequently, it is evident from the above-mentioned facts that amaranth exists as pseudocereals, leaf vegetables, and weeds usually used as ornamental plants. The cultivated species of amaranth have higher nutritional and medicinal values. The amaranth grains are found to be rich in bioactive compounds, such as isoquercetin, nicotiflorin, rutine, ferulic acid, caffeic acid, hydroxybenzoic, and protocatechuic, which possess a diverse range of medicinal properties. For instance, the high content of cysteine and methionine in amaranth protein almost similar to major cereals is used in the preparation of complementary snacks and foods (Escudero et al., 1999). The boiled preparation of roots from amaranth is sued as laxative, treatment of limb pain, tapeworm expellant, and dressing of wounds. Moreover, the squalene found in amaranth are having beneficial effects on cancers and reduction of cholesterol in the blood. It was also reported that this squalene has important beneficial effects on cancers and reduces the cholesterol level in the blood. Amaranth leaves are recommended for patients having low RBC count in East Africa. Regular consumption of amaranth helps in reducing hypertension and cholesterol level and curing cardiovascular diseases.

2.4 Plant-based foods to address the "lifestyle" diseases

Changing scenarios of lifestyle increased the pervasiveness of lifestyle-associated diseases such as diabetes mellitus, hypertension, dyslipidemia, and being overweight that are life threatening to human beings on a global scale. In the proceeding sections, we have discussed the role of foods in regulating lifestyle-associated diseases in humans.

2.5 Role in regulating hypertension

Hypertension is the major risk factor enhancing the clinical features of stroke and cardiovascular diseases, so it is critical to control hypertension to control these debilitating ailments (Lloyd- Jones et al., 2010). Several foods have been introduced to overcome hypertension, which largely includes underutilized crops such as buckwheat (*Fagopyrum esculentum*). The foods based on buckwheat act as a potential antihypertensive food and lead to lower blood pressure (Guang & Phillips, 2009; Zhang et al., 2012). The bioactive compounds obtained from buckwheat result in the reduction of angiotensin I-converting enzyme activity in the lungs, kidneys, thoracic aorta, liver, and heart upon administration in rats (Nakamura et al., 2013). In addition, the "Quercetin" flavonoid found in Tartary buckwheat was employed to treat hypertension by improving insulin sensitivity and reducing oxidative stress (Hou et al., 2017). It is evident from the previous literature that selected food crops can be employed to control hypertension and treat cardiovascular diseases.

2.6 Food having antidiabetic properties

Diabetes mellitus is largely caused by the decrease or loss in insulin synthesis resulting in higher blood sugar levels, that is, hyperglycemia. The disease is further classified into type 1 and type 2 diabetes mellitus. As far as type 1 diabetes is concerned, it is due to damage in pancreatic β cells due to autoimmune diseases. Type 2 diabetes mellitus is due to less production of insulin by pancreatic β cells. Reports suggest that ethanol and rutin extracted from the buckwheat resulted in the suppression of fructose amine and α-dicarbonyl, which in turn reduces protein glycation in diabetic patients (Lee et al., 2015). These compounds improve the uptake of glucose and inhibit the degradation of PPARγ in hepatocytes (Lee et al., 2012). Moreover, the buckwheat hull flavonoids were reported to reduce the weight of db/db mice and white fat index of the abdomen. Tartary buckwheat (TB)-based foods were reported to decrease the negative impacts of type 2 diabetes mellitus and recovery from kidney ailments (Qiu et al., 2016). In addition, it is strongly supported that food based on TB improves health conditions in patients suffering from serum creatinine, urinary albumin to creatinine ratio urea nitrogen, and uric acid abnormalities. All these reports suggest that functional foods based on selected food crops largely control the perseverance of T2DM patients.

2.7 Nutrigenomics, lifestyle diseases, and food quality go hand in hand

The lifestyle diseases are grossly due to the consumption of diversified processed foods and reduction in expenditure of energy due to a sessile or sedentary lifestyle (Popkin, 2006). The influence of nutrients on gene expression is nowadays largely explored by postgenomic approaches. The research field is familiar as "Nutrigenomics" concerned with studying the impact of foods on gene expression and its correlation with the health status of the organism (World Health Organization, 1990). Exploring the complex interaction of nutrients genes and the development of diseases is investigated by the high throughput techniques essential in unraveling the health status and treatment of human ailments. The use of functional foods to treat human diseases is currently a largely explored area to reduce the intake of the drugs with huge side effects. In conclusion, the mentioned applications of nutrigenomics are critical to understanding food interaction on a molecular level and well-identifying food crops with medicinal values.

2.8 Nutrigenomics for improving crop

A large number of disorders of humans have been correlated with the distinctive interactions of nutrients relying on genetic diversity of nutrition and health conditions (Stover, 2006). The know-how of gene−nutrient interaction by employing quantitative techniques in combination with omics techniques. Through microarray techniques, about thousands of genes related to food-related nutrients have been identified. The gene-diet interaction helps to understand the influence of micronutrients and macronutrients on human health. The field of nutrigenomics has helped us to understand the dynamics of nutrients in the cellular process driving the metabolism and central dogma.

2.9 Technical intervention: biofortification of crops through nutrigenomic techniques

The scientific world is still in the infancy of attaining the goal of nutritional and food security through the biofortification of crops by employing the nutrigenomics approach. Nutritional deficiency causes health ailments such as diabetes and obesity in addition to causing serious infections that may prove too fatal (FAO, 2017). Consequently, it is critical to adopt industrial fortification, dietary diversification, and nutraceutical augmentation as remedial measures for nutritional deficiencies (Ruel & Alderman, 2013). In urban areas, the biofortification of crops is the best approach to overcome malnutrition in human populations (Fig. 5.4) (Bouis & Welch, 2010). The qualitative traits, such as the content of starch in wheat (Sestili et al., 2010), protein quality in maize (Gibbon & Larkins, 2005; Yu et al., 2005), and mineral profile in carrots (Morris et al., 2008), can be improved by identifying high-

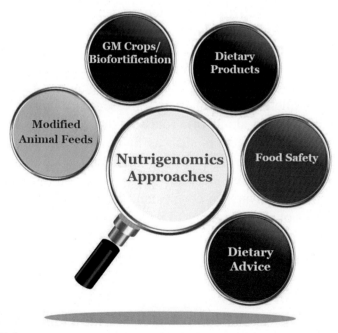

FIGURE 5.4

Technological interventions of nutrigenomic approaches to aid biofortification of crops for nutritional deficiency.

throughput technologies. For instance, the transfer of genes such as coding for natural resistance-associated macrophage protein, iron-regulated transporter, Fe deficiency-specific, nicotianamine synthases, nicotinamide adenine aminotransferase, ferritin, phytase, yellow stripe 1-like transporter, and ferric-chelate reductase oxidase in rice helped ion Zn and Fe biofortification (Masuda et al., 2013; Paul et al., 2014; Trijatmiko et al., 2016). In addition, calcium biofortification is accomplished by the transfer of calcium sensor genes in tomato, bottle gourd, potato, rice, carrot, finger millet, and tobacco and additionally transfer of other genes such as CaMK, TPC, and EcCIPK in finger millets (Han et al., 2009; Mirza et al., 2014; Morris et al., 2008; Vinoth & Ravindhran, 2017). Novel ideas for the biofortification of crops are recently reported as several functional foods and their derivative bioactive compounds have proved therapeutic agents against a large number of diseases. Several high throughput technologies such as SNPs array analysis, NGS, transcriptome profiling, oligo-directed mutagenesis, and gene silencing (RNA interference) (Peng et al., 2016; Schaart et al., 2016). For instance, genome-wide association studies unraveled links between quantitative trait loci (QTLs) of sorghum and rice for polyphenols and iron, respectively (Rhodes et al., 2014). It is reported that high throughput technologies such as TILL-ING/Eco-TILLING, association analysis, and genome editing techniques help in allele mining of traits for essential minerals for crop improvement (Tariq et al., 2014). To assess the qualitative and quantitative traits, several markers such as SNPs, RFLPs, SSRs, STMs, RAPDs, and ESTs have been used to trace critical traits for crop improvement. Marker-assisted selection is a promising approach to develop resilient cultivars. In conclusion, these techniques are front-line approaches of nutrigenomics to improve crops and aid in nutrient biofortification of crops.

3. Conclusion

The nutritional richness of our food plays a critical role in regulating the metabolism and healthy state of the human body. Almost all these nutrients such as proteins, carbohydrates, lipids, vitamins, and minerals are derived from plants and animals. The macro- and micronutrients directly control the metabolic networks in animals and majorly it is their deficiency that is responsible for the development of a wide range of diseases. Consequently, the quality of food essentially plays a critical role in maintaining the healthy state of the body. The profile of nutrition directly influences the expression of a gene important for maintaining health. The field of nutrigenomics is underway to comprehend the relationship between diet and genes to unravel the disease dynamics in humans. Still large and long corroborating studies need to identify all the gene–nutrient interactions through nutrigenomic studies. The efforts of nutrigenomics are grossly dependent on multiple interactions between metabolic profiling, genomics, and environment to unravel the diet–gene interactions. In addition, nutrigenomics deals with the understanding of polygenic diseases related to diet.

References

Almendro, V., & Gascan, P. (2012). *Nutrigenomics and cancer.* http://www.fundacionmhm.org/pdf/Mono9/Articulos/articulo9. pdf.

Balzotti, M. R. B., Thornton, J. N., Maughan, P. J., McClellan, D. A., Stevens, M. R., Jellen, E. N., Fairbanks, D. J., & Coleman, C. E. (2008). Expression and evolutionary relationships of the Chenopodium quinoa 11S seed storage protein. *Gene International Journal of Plant Sciences, 169*(2), 281–291.

Bekkering, C. S., & Tian, L. (2019). Thinking outside of the cereal box: Breeding underutilized (pseudo) cereals for improved human nutrition. *Frontiers in Genetics, 10,* 1289. https://doi.org/10.3389/fgene.2019.01289

Bouis, H. E., & Welch, R. M. (2010). Biofortification—A sustainable agricultural strategy for reducing micronutrient malnutrition in the global south. *Crop Science, 50,* S-20.

Brown, D. C., Cepeda-Cornejo, V., Maughan, P. J., & Jellen, E. N. (2014). Characterization of the Granule bound starch synthase I gene in Chenopodium. *The Plant Genome.* https://doi.org/10.3835/plantgenome2014.09.0051

Corella, D., & Ordovas, J. M. (2009). Nutrigenomics in cardiovascular medicine. *Circulation. Cardiovascular Genetics, 2,* 637–651.

El-Moselhy, M. A., Taye, A., Sharkawi, S. S., El-Sisi, S. F., & Ahmed, A. F. (2011). The antihyperglycemic effect of curcumin in high fat diet fed rats:

The role of TNF and free fatty acids. *Food and Chemical Toxicology, 49*, 1129−1140.

Escudero, N. L., Albarracin, G., Fernandez, S., de Arellano, L. M., & Mucciarelli, S. (1999). Nutrient and antinutrient composition of *Amaranthus muricatus. Plant Foods for Human Nutrition, 54*, 327−336.

FAO. (2017). *The state of food insecurity in the world−meeting the 2016 international hunger targets: Taking stock of uneven progress.* Rome: Food and Agriculture Organization of the United Nations.

Farhud, D. D., Yeganeh, M. Z., & Yeganeh, M. Z. (2010). Nutrigenomics and nutrigenetics. *Iranian Journal of Public Health, 39*(4), 1−14.

Ferguson, L. R. (2009). Nutrigenomics approaches to functional foods. *Journal of the American Dietetic Association, 109*, 452−458.

Gebremariam, M. M., Zarnkow, M., & Becker, T. (2014). Teff (Eragrostistef) as a raw material for malting, brewing and manufacturing of gluten-free foods and beverages: A review. *Journal of Food Science and Technology, 51*, 2881−2895. https://doi.org/10.1007/s13197-012-0745-5

Gibbon, B. C., & Larkins, B. A. (2005). Molecular genetic approaches to developing quality protein maize. *Trends in Genetics, 21*(4), 227−233.

Grancieri, M., Martino, H. S. D., & Gonzalez De Mejia, E. (2019). Chia seed (*Salvia hispanica* L.) as a source of proteins and bioactive peptides with health benefits: A review. *Comprehensive Reviews in Food Science and Food Safety, 18*, 480−499. https://doi.org/10.1111/1541-4337.12423

Guang, C., & Phillips, R. D. (2009). Plant food-derived angiotensin I converting enzyme inhibitory peptides. *Diary of Agricultural and Food Chemistry, 57*(12), 5113−5120. https://doi.org/10.1021/jf900494d

Han, J. S., Park, S., Shigaki, T., Hirschi, K. D., & Kim, C. K. (2009). Improved watermelon quality using bottle gourd rootstock expressing a Ca2+/H+ antiporter. *Molecular Breeding, 24*(3), 201−211.

Hickey, J. M., Chiurugwi, T., Mackay, I., Powell, W., & Implementing Genomic Selection In C.B.P.W.P. Hickey, J. M., et al. (2017). Genomic prediction unifies animal and plant breeding programs to form platforms for biological discovery. *Nature Genetics, 49*, 1297. https://doi.org/10.1038/ng3920

Hou, Z., Hu, Y., Yang, X., & Chen, W. (2017). Antihypertensive impacts of Tartary buckwheat flavonoids by progress of vascular insulin affectability in suddenly hypertensive rodents. *Food and Function, 8*(11), 4217−4228. https://doi.org/10.1039/C7FO00975E

Imamura, T., Isozumi, N., Higashimura, Y., Miyazato, A., Mizukoshi, H., Ohki, S., & Mori, M. (2019). Isolation of amaranthinsynthetase from *Chenopodium quinoa* and construction of an amaranthin production system using suspension-cultured tobacco BY-2 cells. *Plant Biotechnology Journal, 17*, 969–981.

Kelley, N. S., Hubbard, N. E., & Erickson, K. L. (2007). The Conjugated linoleic acid isomers and cancer. *The Journal of Nutrition, 137*, 2599–2607.

Khatibzadeh, S., Micha, M., Afshin, A., Rao, M., Yakoob, M. Y., & Mozaffarian, D. (2012). Major dietary risk factors of chronic diseases: A systematic review of the current evidence for causal effects and effect sizes. *Circulation, 125*, AP060.

Kim, C. Y., Le, T. T., Chen, C., Cheng, J. X., & Kim, K. H. (2011). The curcumin inhibits adipocyte differentiation through modulation of mitotic clonal expansion. *The Journal of Nutritional Biochemistry, 22*, 910–920.

Laviano, A., Rianda, S., Molfino, A., & Fanelli, F. R. (2013). Omega-3 fatty acids present in cancer. *Current Opinion of Clinical Nutrition and Metabolic Care, 16*, 156–161.

Lee, C.-C., Hsu, W.-H., Shen, S.-R., Cheng, Y.-H., & Wu, S.-C. (April 4, 2012). Fagopyrum tataricum *(buckwheat) improved high-glucose-induced insulin resistance in mouse hepatocytes and diabetes in fructose-rich diet-induced mice [research article]. Trial diabetes research.* https://doi.org/10.1155/2012/375673. Hindawi.

Lee, C.-C., Lee, B.-H., & Lai, Y.-J. (2015). Antioxidation and antiglycation of *Fagopyrum tataricum* ethanol remove. *Diary of Food Science and Technology, 52*(2), 1110–1116. https://doi.org/10.1007/s13197-013-1098-4

Lloyd-Jones, D., Lloyd-Jones, D., Brown, T. M., Carnethon, M., Dai, S., De Simone, G., Ferguson, T. B., Ford, E., Furie, K., Gillespie, C., Go, A., Greenlund, K., Haase, N., Hailpern, S., Ho, P. M., Howard, V., Kissela, B., Kittner, S., Lackland, D., … Wylie-Rosett, J. (2010). Leader summary: Heart disease and stroke statistics—2010 update: A report from the American heart association. *Course, 121*(7), 948–954. https://doi.org/10.1161/CIRCULATIONAHA.109.192666

Masuda, H., Kobayashi, T., Ishimaru, Y., Takahashi, M., Aung, M. S., Nakanishi, H., Mori, S., & Nishizawa, N. K. (2013). Iron-biofortification in rice by the introduction of three barley genes participated in mugineic acid biosynthesis with soybean ferritin gene. *Frontiers in Plant Science, 4*, 132.

Mirza, N., Taj, G., Arora, S., & Kumar, A. (2014). Transcriptional expression analysis of genes involved in regulation of calcium translocation and storage in finger millet (*Eleusine coracana* L. Gartn). *Gene, 550*, 171–179.

Morris, J., Hawthorne, K. M., Hotze, T., Abrams, S. A., & Hirschi, K. D. (2008). Nutritional impact of elevated calcium transport activity in carrots. *Proceedings of the National Academy of Sciences of the United States of America, 105*(5), 1431−1435.

Nakamura, K., Naramoto, K., & Koyama, M. (2013). Circulatory strain bringing down impact of matured buckwheat grows in suddenly hypertensive rodents. *Diary of Functional Foods, 5*(1), 406−415. https://doi.org/10.1016/j.jff.2012.11.013

Paul, S., Ali, N., Datta, S. K., & Datta, K. (2014). Development of an iron-enriched high yielding *indica* rice cultivar by introgression of a high-iron trait from transgenic iron-biofortified rice. *Plant Foods for Human Nutrition, 69*, 203−208.

Peng, Y., Hu, Y., Mao, B., Xiang, H., Shao, Y., Pan, Y., & Zhang, G. (2016). Genetic analysis for rice grain quality traits in the YVB stable variant line using RAD-seq. *Mol Genet Genom, 291*(1), 297−307.

Popkin, B. M. (2006). The global nutrition dynamics: The world is shifting rapidly toward a diet linked with non communicable diseases. *American Journal of Clinical Nutrition, 84*, 289−298.

Qiu, J., Li, Z., Qin, Y., Yue, Y., & Liu, Y. (2016). Preventive impact of tartary buckwheat on renal capacity in type 2 diabetics: A randomized controlled preliminary. *Therapeutics and Clinical Risk Management, 12*, 1721−1727. https://doi.org/10.2147/TCRM.S123008

Ramaa, C. S., Shirode, A. R., Mundada, A. S., & Kadam, V. J. (2006). The nutraceuticals—An emerging era in treatment and prevention of cardiovascular diseases. *Current Pharmaceutical Biotechnology, 7*, 15−23.

Rhodes, D. H., Hoffmann, L., Jr., Rooney, W. L., Ramu, P., Morris, G. P., & Kresovich, S. (2014). Genome-wide association study of grain polyphenol concentrations in global sorghum [*Sorghum bicolor* (L,) Moench] germplasm. *Journal of Agricultural and Food Chemistry, 62*, 10916−10927.

Ruel, M. T., & Alderman, H. (2013). Maternal and child nutrition study group. Nutrition-sensitive interventions and programmes: How can they help to accelerate progress in improving maternal and child nutrition. *Lancet, 382*(9891), 536−551.

Schaart, J. G., van de Wiel, C. C. M., Lotz, L. A. P., & Smulders, M. J. M. (2016). Opportunities for products of new plant breeding techniques. *Trends in Plant Science, 21*, 438−449.

Sestili, F., Janni, M., Doherty, A., Botticella, E., D'Ovidio, R., Masci, S., & Lafiandra, D. (2010). Increasing the amylose content of durum wheat through silencing of the SBEIIa genes. *BMC Plant Biology, 10*(1), 144.

Silva-Sánchez, C., De La Rosa, A. P. B., León-Galván, M. F., De Lumen, B. O., De León-Rodríguez, A., & De Mejía, E. G. (2008). Bioactive peptides in amaranth (*Amaranthus hypochondriacus*) seed. *Journal of Agricultural and Food Chemistry, 56*, 1233–1240. https://doi.org/10.1021/jf072911z

Simopoulos, A. P. (2010). Nutrigenetics/nutrigenomics. *Annual Review of Public Health, 31*, 53–68.

Singh, A., & Sharma, S. (2017). Bioactive components and functional properties of biologically activated cereal grains: A bibliographic review. *Critical Reviews in Food Science and Nutrition, 57*, 3051–3071. https://doi.org/10.1080/10408398.2015.1085828

Stover, P. J. (2006). Influence of human genetic variation on nutritional requirements. *The American Journal of Clinical Nutrition, 83*, 436S–442S.

Tariq, A. S., Akram, Z., Shabbir, G., Khan, K. S., Mahmood, T., & Iqbal, M. S. (2014). Heterosis and combining ability evaluation for quality traits in forage sorghum (*Sorghum bicolor* L.). *SABRAO Journal of Breeding & Genetics, 46*(2).

Thiyagarajan, K., Vitali, F., Tolaini, V., Galeffi, P., Cantale, C., & Vikram, P. (2016). Genomic characterization of phenylalanine ammonia lyase gene in buckwheat. *PLoS One, 11*(3), e0151187. Available from: https://doi.org/10.1371/journal.pone.0151187.

Trijatmiko, K. R., Dueñas, C., Tsakirpaloglou, N., Torrizo, L., Arines, F. M., Adeva, C., Balindong, J., Oliva, N., Sapasap, M. V., Borrero, J., Rey, J., Francisco, P., Nelson, A., Nakanishi, H., Lombi, E., Tako, E., Glahn, R. P., Stangoulis, J., Chadha-Mohanty, P., & Slamet-Loedin, I. H., et al. (2016). Biofortified indica rice attains iron and zinc nutrition dietary targets in the field. *Scientific Reports, 6*, 19792. https://doi.org/10.1038/srep19792. 26806528.

Varshney, R. K., Nayak, S. N., May, G. D., & Jackson, S. A. (2009). Next-generation sequencing technologies and their implications for crop genetics and breeding. *Trends in Biotechnology, 27*(9), 522–530.

Vinoth, A., & Ravindhran, R. (2017). Biofortification in millets: A sustainable approach for nutritional security. *Frontiers in Plant Science, 8*, 29.

Vrablík, M., Prusíková, M., Snejdrlova, M., & Zlatohlavek, I. (2009). The omega-3 fatty acids and cardiovascular disease risk: Do weunderstandthe relationship. *Physiological Research, 58*, S19–S26.

World Health Organization. (1990). The Diet, nutrition and prevention of chronic diseases. In *Report of a WHO study group. (WHO technical report series, No. 797), Geneva.*

World Health Organization. (2008). *Interventions on the diet and physical activity.* Geneva: What Works.

World Health Organization. (2011). *Non-communicable diseases.* Geneva: Country and Profile.

Yasui, Y., Wang, Y., Ohnishi, O., & Campbell, C. G. (2004). Amplified fragment length polymorphism linkage analysis of common buckwheat (*Fagopyrum esculentum*) and its wild self-pollinated relative *Fagopyrumhomotropicum. Genome, 47*, 345−351.

Yu, J., Peng, P., Zhang, X., Zhao, Q., Zhu, D., Sun, X., & Ao, G. (2005). Seed specific expression of the lysine-rich protein gene sb401 significantly increases both lysine and total protein content in maize seeds. *Molecular Breeding, 14*, 1−7.

Zhang, Z.-L., Zhou, M.-L., Tang, Y., Li, F.-L., Tang, Y.-X., Shao, J.-R., Xue, W.-T., & Wu, Y.-M. (2012). Bioactive mixes in useful buckwheat food. *Food Research International, 49*(1), 389−395. https://doi.org/10.1016/j.foodres.2012.07.035

Analysis of proteomes—I

1. Polyacrylamide gel electrophoresis and SDS-PAGE

The gel-based techniques are very important for the separation and quantification of biomolecules and essential for carrying high throughput mass spectrometric analysis of analytes. These approaches help to resolve thousands of protein isoforms under controlled modulations. The sample to be resolved may be from any source such as homogenized tissue, bacterial sample, or some clinical sample. The technique is generally used for separating protein and DNA samples from the lysed material of cells. The method helps to resolve the components of the protein mixture on the basis of size and shape. In this technique, a charged molecule is under the influence of the electric current. The general polyacrylamide gel electrophoresis (PAGE) is not able to determine the molecular weight of protein samples due to the dependency of mobility on size and charge.

The techniques aid in the separation of biomolecules such as proteins critically based on their mobility under the influence of electricity in an inert gel. The mobility of proteins is largely dependent on their size, charge, and shape. Which in turn depends on the composition of amino acids and the type of secondary structures they form. The frequently used gelling agent polyacrylamide leads to the formation of small pores upon polymerization.

1.1 Principle of PAGE

The proteins are isolated through pores formed as a result of polyacrylamide gel under the influence of electric current. The pores are formed by using a mixture of acrylamide—bisacrylamide polymerized

Principles of Genomics and Proteomics. https://doi.org/10.1016/B978-0-323-99045-5.00004-5

Table 6.1 Percentage of acrylamide used for forming gels to resolve proteins of different molecular weights.

Percentage of acrylamide	Separating resolution of proteins (MW)
5%	60–220 KDa
7.5%	30–120 KDa
10%	20–75 KDa
12%	17–65 KDa
15%	15–45 KDa
17.5%	12–30 KDa

by the action of ammonium persulfate (APS). The polymerization reaction is catalyzed by tetramethylethylenediamine (TEMED) to form a net-like architecture. The pore size of the gel depends upon the percentage of total acrylamide gel. In addition to this, the ration of acrylamide to bisacrylamide also influences the pore size (Table 6.1). Smaller pore size impedes the mobility of proteins and at the same time enhances the resolution and prevents running off from the gel under the influence of higher voltage.

Sodium dodecyl sulfate (SDS)—polyacrylamide gel electrophoresis is the most widely used technique used to resolve the mixture of proteins on the basis of size for the determination of relative molecular mass. The SDS is an anionic detergent that imparts a negative charge on the denatured polypeptide chain. Before running protein mixtures in the SDS-PAGE, the protein samples are boiled in a buffer containing SDS and β-mercaptoethanol for 5 min. Disulfide bonds between proteins are reduced by β-mercaptoethanol. This treatment opens up the proteins into a rod-like primary structure. It is reported that 01 SDS molecule interacts with at least 02 amino acid residues, hence resulting in dampening the charges in the native proteins thus denaturation.

The sharpening of bands is obtained by the formation of glycinate ions in the buffer because the protein—SDS complex has a higher molecular weight than glycinate ions. The chloride ions in the buffer move at the same speed as the glycinate ions. The protein—SDS

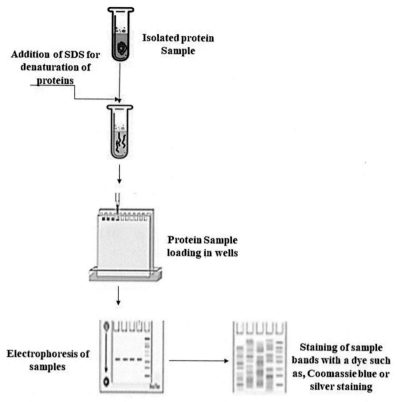

FIGURE 6.1

Schematic diagram showing the SDS—polyacrylamide gel electrophoresis (SDS—PAGE)-based gel electrophoresis of protein samples.

complex runs in between chloride and glycinate ions. During electrophoresis, the protein—SDS complex moves toward the anode depending on the size of the polypeptide chain. As during the process of electrophoresis, the bromophenol blue dye moves faster because it is not retarded by the sieve effect..

1.2 Basic requirements and workflow of PAGE (Fig. 6.1)
Equipment's required:

- An electrophoresis tank
- Electrophoresis frame

- Power supply-power pack
- Glass plates
- Comb
- Spacers
- Casting frame

1.3 Buffers used in PAGE

✔ **Gel casting buffer**—for preparation of gel
✔ **Sample buffer**—buffer used for sample preparation
✔ **Running buffer**—used for electrophoresis of the protein

In general, tris-based buffers are utilized in PAGE analysis, for instance, bis-tris, Tris-glycine, tris-tricine, and tris-acetate are frequently used tris-based buffers. The tris-borate-ethylenediaminetetraacetic acid (TBE) is the most frequently used buffer in native PAGE.

1.4 The gel

The varying percentage of polyacrylamide is for forming the gel and appropriate pore size (refer Table 6.1). The available form of acrylamide is usually supplied in form of liquid and is found to be neurotoxic. As already explained, polymerization is achieved by mixing the bis-acrylamide and acrylamide formed between acrylamide units. Acrylamide is used due to several salient features such as large samples being relatively accepted, resolving power being high for moderate- and small-size proteins and nucleic acids, very less interaction with sample, and higher stability of matrix formed.

1.4.1 Polymerization of polyacrylamide gel

To initiate polymerization, free radicals are generated by the addition of APS and by the addition of TEMED. TEMED is used as a successful catalyst and aids in the decomposition of APS to generate freer radicals as shown in the following reaction:

$$S_2O_8^{2-} + e^- \rightarrow SO_4^{2-} + SO_4^{-\cdot}$$

The polyacrylamide gels can alternatively be polymerized by photopolymerization. Riboflavin may also be sued for induced

Table 6.2 Composition of resolving and stacking gels used for resolving the proteins.

Resolving gel (10%)	
Distilled water	4.0 mL
30% acrylamide mix	3.3 mL
1.5M Tris (pH8.8)	2.5 mL
10% sodium dodecyl sulfate (SDS)	0.1 mL
10% ammonium persulfate (APS)	0.1 mL
Tetramethylethylenediamine (TEMED)	0.004 mL
Stacking gel (5%)	
Distilled water	5.65 mL
30% acrylamide	1.65 mL
1.0M Tris (pH 6.8)	2.5 mL
10% sodium dodecyl sulfate (SDS)	0.1 mL
10% ammonium persulfate (APS)	0.1 mL
Tetramethylethylenediamine (TEMED)	0.004 mL

photochemical polymerization in the presence of UV radiation. These chemicals are mixed in a glass beaker, once polymerization begins the mixture is loaded between the assembled glass plates for gel formation.

The prepared gel is divided into two different types of gels, viz., stacking gel having a larger pore size at 5% acrylamide and resolving gel with a smaller pore size for higher resolution (Table 6.2). The composition and pH of both the gels vary, for stacking gel pH 6.8 to concentrate protein samples such that sharp bands will be formed in the resolving gel with 10% acrylamide at a set pH of 8.8. The chloride ions are mobile in both gels.

1.5 Preparation of protein samples

The preparation of samples varies depending on whether using PAGE or SDS-PAGE. The sample is mixed with ionizable tracking dye such as bromophenol blue for monitoring the electrophoretic run. In addition, glycerol or sucrose for maintaining the density of the protein sample such that sample will be easily settled in the wells during loading. The SDS, β-mercaptoethanol, glycerol, and bromophenol blue are boiled at 100°C for 3 min. Before loading, the samples are centrifuged to remove the debris.

1.6 Running the gel

The electrophoresis tank is filled with running buffer (TBE in case of native PAGE and tris-glycine SDS). Do not remove the comb from the casted gel to maintain the integrity of the wells until set in the electrophoresis tank. Remove the comb carefully and load samples using a 20 μL micropipette. During the loading of the sample, be careful not to damage the wells. After loading the sample and the molecular weight marker ladder, the appropriate voltage is applied. Voltage depends on the set up such as PAGE or SDS-PAGE. Generally, 100 V for 40 min in the case of native PAGE and 200 V for 35 min for SDS-PAGE is usually applied. Once electrophoresis starts, the glycine in the buffer is passed into stacking gel lagging behind chloride ions.

1.7 Staining of gel and visualization of proteins

The gel is stained by Coomassie Brilliant Blue 01 h for a typical mini gel. Overnight destaining of the gel is important for removing the stain completely from the gel. Once the sample has sufficiently run through the gel, the gel frame is removed from the tank and carefully removed from the glass plates. The stacking part of the gel is cut to leave the resolving gel for developing a clear image of proteins. The resolving part of the gel is subjected to destaining in a solution containing methanol 20% (v/v), glacial acetic acid 10% (v/v), and Coomassie brilliant blue 0.1% (w/v) at least for 01 h. The destaining process can also be obtained by leaving gels in the destaining solution overnight. Alternatively, silver staining can also be used for certain specific purposes. The protein bands in the gel are visualized by the naked eye and the image can be captured by high-resolution cameras.

1.8 Gel interpretation

The amount of protein is detected by the strength of the band and comparison with the standard ladder.

2. Isoelectric focusing gels

This method is employed for the separation of amphoteric molecules such as proteins on the basis of varying isoelectric points. This method can separate proteins that differ in very less isoelectric points. This technique utilizes glass plates or horizontal gels or plastic sheets about 12 mm thick with an established pH gradient. The pH gradient in the gels is generated by the introduction of ampholytes derived from synthetic poly-amino-polycarboxylic acids such as Pharmalyte and Bio-Lyte. Usually, isoelectric focusing gels (IEF) are carried at a low percentage of polyacrylamide (4%).

2.1 Preparation of IEF gels

Thin layer IEF gels are prepared by carrier ampholytes mixed with riboflavin and acrylamide solution and then poured over the glass plate with spacers. Now the second plate is placed over the plate first to form a gel cassette and the polymerization is obtained by exposing the plates to bright light (photopolymerization) at least for 23 h. The decomposition of riboflavin by bright light leads to free radicals critical for intuiting the process of acrylamide polymerization. After the gel is solidified and set, the plates are separated and the gel stuck plate is processed for the further technical process. The electrode wicks were laid along the longer length of the gel and subsequently, voltage is applied. Under the influence of potential difference, the ampholytes establish a pH gradient between cathode and anode.

   After the establishment of the pH gradient, the power supply is switched off and the protein samples are laid on the gel in small squares placed on small filter paper. After placing the samples, the voltage is again applied for about 30 min. Once run the protein samples the pH steadily increases and the isoelectric points are set. For obtaining rapid separation of proteins higher voltage of about 2500 V up to 23 h can be applied. After completion of electrophoresis, the gel is stained to visualize the protein spots. Coomassie Brilliant Blue is sued to stain the gel and then subsequently destained from the identification of proteins. The distance of the protein bands is measured from one electrode from each point against the pIs.

IEF is used to study the microheterogeneity in a protein in contrast to SDS PAGE. The technique is also useful for studying isoenzymes that usually differ in very few amino acids. Moreover, the technique is useful to investigate protein profiles in forensic science. Apart from being useful in the analytical separation of proteins, IEF is also sued for preparative purposes of biomolecules.

3. 2D gel electrophoresis-combining SDS-PAGE and IEF for unison gel-based identification of proteins on the basis of size and charge

The techniques separate the proteins on the basis of pH and molecular mass as mentioned earlier for better precision and resolution from a mixture of proteins isolated from the cells, tissues, and other biological fluids (Klose, 1975; O'Farrell, 1975). The 2D-GE helps to resolve up to 5000 protein spots with high precision and separation in one go (Magdeldin et al., 2012). The technique provides a platform for further analysis of differently regulated protein samples by western blotting and mass spectrometry.

3.1 Basic principle of 2D-GE

Two basic technical interventions are required in this technique, viz., first and second dimensions. As far as the first dimension is concerned, the proteins are resolved by using IEF as explained in the previous section (Magdeldin et al., 2012). Various tactics, such as IEF, immobilized gradient electrophoresis, and nonequilibrium pH gradient electrophoresis (NEPHGE), are used to separate the proteins under the pH gradient. With respect to the second dimension, the proteins are separated on the basis of MW by suing tris-tricine or SDS-Laemmli, hence leading to the efficient separation of proteins with similar physicochemical properties (Magdeldin et al., 2012).

3.2 Procedure for setup the 2DGE

3.2.1 Preparation of gel for one-dimension IEF process

✔ The gel tubes are marked up to 1—3 mm for the desired length of the gel. The dimensions must same as that of the second dimension.

↙ The rubber band was placed around the gel tubes around 12 tubes to form a bundle.

↙ One end of the tube is carefully sealed by using parafilm for tight sealing.

↙ The bundle of gel tubes is placed inside the gel casting tube in a vertical position.

↙ Standard amounts of water, urea, acrylamide, bisacrylamide, and ampholytes are added to a small vacuum flask. The mixture was stirred by immersing magnetic stirrers until all the contents are completely dissolved.

↙ The solution was then filled in a 20 mL syringe fitted with filter capsules and passed through the filter.

↙ TEMED and ammonium persulfate were added to the above solution, and immediately the solution was loaded into gel tubes

↙ The parafilm was removed from the bottom of the casting tube, and the excess of the acrylamide was cut by a razor blade.

↙ The air bubbles were removed from the gel tubes.

↙ Power supply was applied to the tube cells. About 200 V for 1 h and then switched off to proceed with sample loading.

↙ A protein layer protein (100−150 μg) was loaded on the gel by using a 50 μL microsyringe.

↙ The lid was placed over the upper reservoir, and the electrical wires are connected for applying electric current.

↙ The power supply with a voltage of 700−800 V for 16 h was switched ON for resolving proteins on the basis of pIs.

↙ The power supply was switched OFF, and about 1 μL of bromophenol blue was added on the top of the gel layer by using a microsyringe.

↙ The gel was extruded from gel tubes by water pressure generated by a syringe.

↙ The gels were stored at −70°C for several weeks or immediately processed for the second dimension (SDS-PAGE).

3.2.2 Preparing and pouring of gel for second-dimensional gels (SDS-PAGE)

↙ The gel plates were assembled, and the clamps were positioned on the sides of the plates.

✔ Level and vertical adjust the setup.

✔ The gel solution was prepared by adding gel buffer and 30% acrylamide/0.8% bisacrylamide to water in a vacuum flask, and vacuum was applied to the solution for up to 5 min.

✔ To the above solution, add TEMED and 10% SDS, mix and swirl the gel solution and immediately add APS.

✔ The above gel solution was poured into plates and overlayed with water.

✔ The gel was allowed to polymerize for >1.5 h.

3.2.3 Fitting and loading first-dimension gel (IEF) onto second-dimension gel (SDS-PAGE)

✔ The first-dimension gel was sided and placed over the top of the gel slab.

✔ Voltage was applied, and the gel was stained and destained for identifying the desired spots of proteins as discussed in the previous section.

4. MALDI-TOF mass spectrometry

The techniques of mass spectrometry were used to read the molecular weight of biomolecules based on mass-to-charge (m/z) upon ionization by lasers. The ion source in MALDI-TOF is a matrix of diverse classes. The basic principle involves striking of lasers on small molecules to convert them into gas phase without decomposing or fragmenting the biomolecules such as proteins, peptides, lipids, carbohydrates, and the vast majority of other organic molecules..

4.1 The principle and setup of MALDI-ToF (Fig. 6.2)

In this technique, the analyte is embedded into the matrix compound usually embedded on the solid support made of conducting metal and has a large number of spots for loading the sample analyte. After the laser is directed on the sample spots, the analyte is rapidly heated and becomes vibrationally excited. The laser energy is absorbed by the matrix molecules, which in turn carry the analyte into the gas phase for travel in the flight tube. During the analyte ablation, the analyte molecules are ionized rather than deprotonated or protonated.

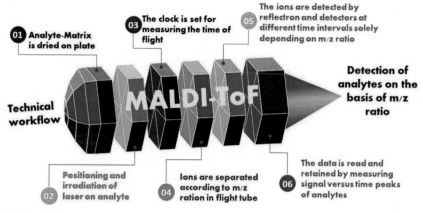

FIGURE 6.2

Workflow of MALDI ToF for detecting the protein fragments on the basis of m/z ratio.

4.2 Basic instrument and working setup

4.2.1 Sample concentration for MALDI

The samples are diluted properly at a particular concentration. Proteins and peptides at $0.1-10$ pmol mm^{-3} concentration and polymers around 10 pmol mm^{-3} give the best spectra. The analyte is dissolved in 0.1 mg/mL in the desired solvent. To this solution, 10 mg/mL of the matrix is added to yield a concentrated or saturated solution for proper analysis. The optimum molar matrix-to-analyte ratio ranges from 1000:1 to 100,000:1. After preparation, the analyte is loaded on spots on the plate. After loading and proper drying, the mixtures and matrix compounds crystallize, and subsequently, the plate is fit in the MALDI-ToF instrument setup.

4.2.2 Sampling plates used in MALDI-ToF

- **Stainless steel flat plates (100-well)**
  In these plates, external calibration is sued and the crystallization is easily seen on the matrix surface. At least 100 analyte samples can be loaded for m/z analysis.
- **Teflon-coated plates (400 spots for samples):**
  This type of plate enhances the sensitivity of samples. The sampling is automated to increase the accuracy of spots.

- **Gold-coated plates**

  These plates contain a 2 mm diameter to properly spread the matrix and analyte. The well allows the reaction between gold surfaces and thiol-containing reagents.

4.3 The most commonly used matrix compounds in MALDI ToF (Table 6.3)

The isolation and dilution of analyte is the primary purpose of matrix compounds embedded on the plates. Matrix aids mediator for absorbing the energy of lasers. Consequently, it is critical to select the right matrix for the right type of analyte. The most commonly used matrix includes 3,5-dimethoxy-4-hydroxycinnamic acid, 2,6-dihydroxyacetophenone, and α-cyano-4-hydroxycinnamic acid.

4.4 Laser types used in MALDI-ToF

In general, infrared and ultraviolet lasers have been sued to activate the analytes loaded on the sampling matrix plate. The following lasers have been used in MALDI ToF:

- Nitrogen lasers (337 nm)
- Frequency-tripled Nd: Yag lasers (355 nm)
- Quadrupled Nd: Yag lasers
- Er: Yag lasers
- TEA−CO_2 lasers

Upon laser shooting, the matrix as well as analyte sublimates and ionizes, and the analyte ions are generated and desorption occurs through the ToF analyzer on the basis of mass/charge (m/z) ratio.

4.5 Liner TOF analyzer and reflectron TOF analyzer

The laser pulse is not striking the analyte at the same time, so the velocities of analytes vary even though they have the same m/z ratio. To correct this error, the reflectron is set and consists of a series of electrodes operating on high voltage to repulse the ions back to the integrated flight tube. During the process of analyte flight, the faster ions travel faster due to higher kinetic energy than slower ones. Hence,

Table 6.3 Matrix used in resolving a diverse range of analytes in MALDI-ToF.

S. No.	Name of the matrix	Analyte resolved
1.	3-Aminopicolinic acid (3-APA)	Oligonucleotides, DNA
2.	6-Aza-2-thiothymine (ATT)	Oligonucleotides, DNA
3.	3-Hydroxypicolinic acid (HPA, 3-HPA)	Oligonucleotides, DNA
4.	2,5-Dihydroxybenzoic acid (DHB)	Proteins, oligosaccharides
5.	3-Aminoquinoline (3-AQ)	Oligosaccharides
6.	α-Cyano-4-hydroxycinnamic acid (α-CHC, α-CHCA, 4-HCCA, CHCA)	Peptides, smaller proteins, triacylglycerols, and numerous other compounds
7.	4-Chloro-α-cyano-cinnamic acid (ClCCA)	Peptides
8.	3,5-Dimethoxy-4-hydroxycinnamic acid (SA)	Proteins
9.	2-(4-Hydroxyphenylazo) benzoic acid (HABA)	Peptides, proteins, glycoproteins, polystyrene
10.	2-Mercaptobenzothiazole (MBT)	Peptides, proteins, synthetic polymers
11.	5-Chloro-2-mercaptobenzothiazole (CMBT)	Glycopeptides, phosphopeptides, and proteins
12.	2,6-Dihydroxyacetophenone (DHAP)	Glycopeptides, phosphopeptides, proteins
13.	2,4,6-Trihydroxyacetophenone (THAP)	Solid-supported oligonucleotides
14.	9-Nitroanthracene (9-NA)	Fullerenes and derivatives
15.	2-[(2E)-3-(4-tert-butylphenyl)-2-methylprop-2-enylidene] malonitrile (DCTB)	Oligomers, polymers, dendrimers, small molecules
16.	Benzo[a]pyrene	Fullerenes and derivatives
17.	Dithranol (1,8,9-anthracenetriol)	Synthetic polymers
18.	DHB-based mixtures (DHB/XY and super-DHB)	Proteins, oligosaccharides
19.	Nicotinic acid (NA)	Peptides, proteins
20.	Picolinic acid (PA)	Oligonucleotides, DNA

upon detection, the resolution of activated biomolecules is enhanced considerably.

4.6 The reflectron for higher resolution

The reflectron aids in increasing the path length of the flighted ions and rectifies the little variation. The reflectron system has established an electric field gradient in which the ions will penetrate. Hence, overall the ion flight times become focused. In conclusion, the reflectron enhances the overall mass accuracy, resolution, and information regarding sequence information and structure.

4.7 Applications of MALDI ToF

4.7.1 Investigation of proteomes

The technique provides a quick investigation of the primary structure of proteins in a few minutes. The higher levels of proteins can be examined by using rapifleX and ultrafleXtreme instrumental setup of MALDI-ToF.

4.7.2 Identification of microorganisms

The microorganism can be directly processed through MALDI-ToF without any chemical pretreatment. Since vegetative bacteria can be easily lysed by exposure to organic solvents, strong acids, or even water found in the matrix. Even though pretreatment is essential for bacterial species such as Actinomyces. Two approaches are utilized to identify and characterize the microorganism, viz., comparison of mass spectra with already fingerprints in databases and masses of microbial biomarkers with the existing proteome database.

4.7.3 Discovery of biomarkers

The technique works directly with tissues to track the molecular changes originating at the cellular level or in solutions.

4.7.4 MALDI for bioimaging

The technique aids in the rapid collection of data sets and helps in maintaining essential spatial data and the development of precise tissue typing.

4.7.5 Characterization of polymers

Screening of polymers from pharmaceutical products and bulk materials helps in versatile and fast identifications of polymers.

4.8 Quality control

Due to the variety in ionization matrices, the technique is critical for assessing the quality control of oligonucleotides, peptides, and other biopolymers.

5. Isotope-coded affinity tags

It is a technique in which the protein samples are labeled by chemical reagents for quantitative estimation of proteomes to analyze through mass spectrometry (Colangelo & Williams, 2006; Gygi et al., 1999). The labeled probes consist of the following:

- A reactive group such as iodoacetamide or *N*-ethymaleimide tagged to the amino acid side chain. The chemical reactive group alkylates the cysteine thiol group in protein samples.
- Isotope coded linker or linker arm meant for mass shift introduction of different samples of peptides. Isotopes, such as proton (H)/ deuterium (D), 12C/13C, and 15N/16N, are used for obtaining residues of light and heavy tags.
- A tag such as biotin is used for purification of the labeled peptides through affinity chromatography (biotin-avidin/streptavidin systems) to capture all the cysteine-containing peptides from the mixture.

All the protein samples are mixed and combined and subsequently separated by chromatography and then mass spectrometry for the mass-to-charge determination of protein samples. The limitation for this technique is that only cysteine-containing amino acids are analyzed.

5.1 Basic principle and workflow (Fig. 6.3)

The basic principle involves protein isolation from two different samples followed by labeling with isotope-coded affinity tag (ICAT)

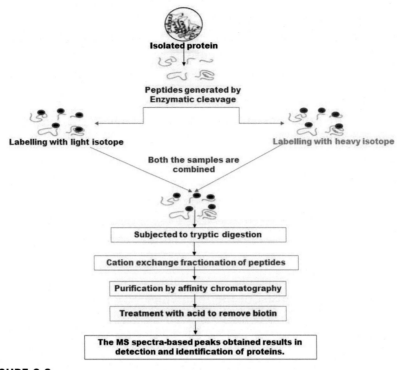

FIGURE 6.3

Workflow for spectral analysis of proteomes by ICAT proteomic technique.

reagents with heavy and light isotopes. The samples are pooled and subjected to enzymatic digestion. The samples are then purified by affinity chromatography by biotin for reduction of complexity. Subsequently, the biotin is removed to enhance the resolution through the use of mass spectrometry by photo-cleavable linkers (UV light) or acid-cleavable (ALICE/introduction of a disulfide bond in linker) (Bottari et al., 2004; Qiu et al., 2002). Further, the resolved samples are subjected to liquid chromatographic separations and tandem mass spectrometric techniques. The MS spectra-based peaks obtained results in the detection and identification of proteins. In conclusion, the technique offers high accuracy and reduces complexity, even though the technique leads to loss of information.

5.2 Applications of isotope-coded affinity tag

✔ It is applied to measure the pairwise changes in the expression of proteins through the use of differential labeled isotopes and their subsequent identification and quantification of proteomes.

✔ **Absolute quantification of proteomes:**

The techniques exploit the use of standard synthetic tags to determine the femtomole levels of proteins in biological samples of interest. The method is applicable to closely related protein isoforms, hydrophobic proteins. Protein such as cytochromes P450 (P450) and metabolic enzymes, are readily determined by the ICAT approach.

✔ **Identifying and quantifying oxidant-sensitive protein:** the approach is employed to qualify and identify the oxidant-sensitive protein thiols by employing.

✔ The technique is used for quantitative protein profiling of a diverse range of living samples.

✔ The technique aids in the quantitative profiling of cancer cells in combination with mass spectrometry.

✔ Proteomic analysis based on 2-DE/MS and cleavable ICAT helps to determine aromatic compounds involved in a range of metabolic pathways in living systems.

6. Isotope-coded protein labeling

This technique is a modified version of ICAT to resolve the limitations associated with higher sequence coverage (Schmidt et al., 2005). The technique uses *N*-nicotinoyloxy-succinimide tags that cause the derivatization of free amino-terminal groups and ε-amino groups of lysine residues. The technique introduces mass differences of 4 and ∼6 Da in the case of H/D and 12C6/13C6 labeling. The steps include extraction of protein, reduction in protein, and alkylation for ensuring the uniform labeling of lysine amino acids. The heavy and light labeled peptides are pooled and subjected to digestion with enzymes. In addition to trypsin, the peptides are digested with Glu-C endo-proteinases to obtain shorter fragments. The tags used in this technique are hydrophilic for efficient quantification. Varied combinations of isotope-coded protein labeling (ICPL) tags are used. The technique helps to permit lysine labeling rather than cysteine to extend the coverage of proteins. All the N terminal peptides fragments are labeled uniformly to make sure accurate and nonbiased results are obtained (Leroy et al., 2010). Subsequently, the labeled fragments are separated by liquid chromatography. The separated fragments are analyzed by MS to quantify and identify varying fragments of proteins. The technique helps to efficiently detect the proteins, and it also aids in the analysis of isoforms and posttranslational modifications in proteins (Lottspeich & Kellermann, 2011). The technique is employed to quantify proteins with small differences of even 1 Da.

6.1 Applications

- The technique aids in the quantitative proteomic analysis as well as a bioinformatic analysis-based examination of vitreous protein contents and their associated pathways.
- The ICPL is high throughput technic employed for quantitative proteome profiling of thousands of proteins.
- The ICPL is used to assess the differential proteomes through primary labeling of amines of proteins.

7. Isobaric tags for relative and absolute quantification

The technique is employed to explore a wide range of analytical data linked to biomolecules. This technique involves isobaric labeling of peptides to N-terminals through a multiplexing approach. The labeled tag consists of three components, viz.,

- Reporter group—for mass shift, such as N-methylpiperizine. The mass ranges from 114 to 117 Da and 113 to 121 Da.
- Balance group—to maintain the mass of isobaric tag such as carbonyl group. The mass of balance ranges from 28 to 31 Da and 184 to 192 Da in 4-plex (mass tag = 145 Da) and 8-plex (mass tag = 305 Da).
- NHS-binds to specific peptides such as amine-reactive ester group.

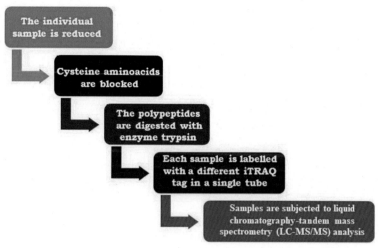

FIGURE 6.4

Workflow for analyzing the protein samples through the iTRAQ technique.

The techniques help to compare 02, 04, and 08 samples simultaneously for efficient comparisons. In this technique, the digested peptides labelled with the NHS groups are analyzed by LC-MS/MS. Additionally, the labeled peptides are resolved by separating through chromatographic techniques and detection by MS to generate appropriate spectra. In the second round of fragmentation, their labeling and detection are important to resolve the missing peptides. It is pertinent to use high-quality triple quadrupole MS analysis.

7.1 Basic principle and workflow (Fig. 6.4)

The peptides covalently linked with isobaric tags for relative and absolute quantification (iTRAQ) reagent isobaric tags at the lysine side chain of each peptide are subjected to the analysis. During the process, the mixture of peptides is labeled with different iTRAQ reagents and subsequently labeled to proceed for tandem mass spectrometry. As a result, the mass spectrum obtained yields similar peaks due to identical molecular weight even though the same peptides are obtained from different sources. Henceforth, the peaks of the same attribute are subjected to second-stage mass spectrometry. This step is obtained by breaking the bond between the peptide reactive group and the balance group leading to the loss of the balance group. This step leads to production of different isotope labels with different masses. The data is then analyzed by softwares and existing databases.

7.2 Some advantages of iTRAQ

✔ Precise accomplishment of quantitative proteomics using mass spectrometry.
✔ The metabolic progression of drugs can be monitored.
✔ The quantification of biomarkers at different stages of the disease can be detected.
✔ The ability of multiplexing samples.
✔ The time and variation are reduced considerably.

7.3 Some disadvantages of iTRAQ

✔ The reagents in iTRAQ are extremely costly and highly sensitive to contamination with salts.
✔ The inefficiency in enzyme-based digestion of proteins leads to variable data sets.
✔ Highly erudite softwares are used to analyze the iTRAQ-generated data.

7.4 Applications

✔ The technique has been employed to identify the diseases such as idiopathic pulmonary fibrosis, and in addition, the technique is aided in unraveling novel noninvasive biomarkers for diagnostics.
✔ Identification of potential markers for cancers such as primary pulmonary adenocarcinoma has been accomplished by studying the abnormal posttranslational protein modifications critical for the progression of lung cancers.
✔ The technique has been used to monitor the proteomes of neurological disorders associated with proteomic changes.
✔ It is widely used for investigating signaling pathways.

8. The tandem mass tag

The technique is based on a similar principle as that of iTRAQ with very slight modification in the chemical nature of tags used for labeing peptides. The mass difference is created by using 13C/15N isotopomers. The tag consists of a linker region, a balance group (with 99–104 Da which leads to 6-plex analysis), a reporter group (varies from 126 to 131 Da), and a reactive group that usually binds to amine or cystine or carbonyl groups of amino acids (Dayon et al., 2008). The technique also leads to uniform labeling of fragments at amines or cysteine residues obtained after the enzymatic digestion of proteins. The release of reporter is accomplished at the time of peptide fragmentation during MS/MS analysis. This step helps in yielding fine spectra for the relative identification and quantification of proteomes. The multiplexing approach aids in the proficient comparison of proteins with complicated data. The biasness and inaccuracy due to the pooling of samples before LC-MS/MS analysis is a major drawback associated with tandem mass tag. The limitation is rectified by employing metabolic labeling, which involves the biological incorporation of amino acid isotopes and other elements through the intervention of cell cultures.

9. Stable isotopic labeling of amino acids in cell culture

The stable isotopic labeling of amino acids in cell culture (SILAC) in an in vivo proteomic technique was first time developed by Ong et al. (2002). The process of SILAC involves the direct addition of stable isotope amino acids in cell culture to investigate the proteomes of cells under investigation. The technique has straightforward application in accurate quantitation and higher reproducibility over methods such as chemical labeling or label-free quantitative methods. Consequently, this method is widely used for the characterization of proteomes in a wide range of biological samples for critical investigation of posttranslational modifications (PTMs). The technique involves isotopic lysine ($C_6H_{14}N_2O_2$) and arginine ($C_6H_{14}N_4O_2$) amino acid tags for the detection of proteomes in cells. The technique is simple and involves preparation of culturing of cells to be investigated for proteomes of interest.

9.1 Principle of SILAC

The basic principle behind SILAC involves the metabolic incorporation of stable isotope-labeled amino acids into the proteome of the cell during the culturing process. In this technique, the SILAC 2 cell populations are grown in two separate culture mediums, one containing light medium isotope-labeled amino acids and the other culture containing heavy medium stable labeled amino acids. After five cycles of cell division, it is assumed that all the cells in the culture will be labeled with a heavy medium containing amino acids (Mann et al., 2013), which grossly depends on translation, degradation, and protein turnover. Further, cell populations are experimentally manipulated and subsequently the unlabelled and labeled cells as well as protein extracts are mixed thoroughly. After this step, the samples are digested into smaller peptides by using trypsin and the proteins are separated by suing SDS-PAGE. These peptides are then analyzed by employing LC-MS/MS. Finally, the ration of labeled and unlabelled peptides is used for the quantification of proteins. The posttranslational modifications, methylation in particular, is analyzed by employing the SILAC approach (Ong et al., 2004). The amino acids sued in the SILAC technique are usually essential amino acids such as leucine (Ong et al., 2002), lysine, and methionine (Ong et al., 2004).

9.2 The technical workflow of SILAC

The workflow comprises two major steps, viz, an adaptation phase and an experimental phase. The technique uses dialyzed serum for culturing of cells such that accumulation of free amino acids will be avoided (Gehrmann et al., 2004). In addition, a very less amount of growth factors is added to the culture medium for proper growth (Cui et al., 2009; Gehrmann et al., 2004). During adaptation phase, the cells are grown in both labeled and unlabelled medium until all the heavy amino acids are incorporated in the cells. At least five subcultures are needed to incorporate cells with heavy isotope-labeled amino acids. The amino acids are then analyzed by LC-MS/MS.

The two populations of cells are finally treated with two different reactions followed by cell lysis, organelle purification protein extraction, and purification and digestion. Subsequently, the protein samples are analyzed by LC-MS/MS for identification and quantification of heavy as well as light-labeled amino acids in peptides. The SILAC analysis is generally performed by using Orbitrap-based mass spectrometers such as quadrupole Orbitrap (Q-Exactive) or by linear ion-trap Orbitrap (LTQ-OrbitrapVelos) (Kelstrup et al., 2012; Michalski et al., 2011; Olsen et al., 2006). Further, the high-quality MS data are processed for the identification and quantification of peptides by suing databases like Census (Park et al., 2012), pQuant (Liu et al., 2014), MaxQuant (Cox & Mann, 2008; Tyanova et al., 2014), and trans-proteomic pipeline (Keller & Shteynberg, 2010). Among these databases, the MaxQuant is frequently used for the analysis of SILAC data. To convert the processed data into meaningful results and biological insights, the data are subjected to analysis through bioinformatics tools such as DAVID (Huang da et al., 2009), cytoscape (Shannon et al., 2003), and GoMiner (Jensen et al., 2009) or annotation databases such as KEGG (Kanehisa et al., 2004) and GO (Ashburner et al., 2000) (Fig. 6.5).

9.3 SILAC-based applications

9.3.1 In analysis of changing dynamics in proteome expression

The technique is used to compare the expression of proteomes during cell differentiation like in the case of adipocyte differentiation (Lossner et al., 2011) and muscle cell differentiation (Cui et al., 2009; Ong et al., 2002), to assess the differential expression of proteins.

9.4 Studies of secretomes

One of the interesting applications of SILAC is the study of the secretome, which is the sum of all proteins released into the extracellular environment by a specific cell or cell type. Proteins such as growth factors, cytokines, hormones, and cancer proteins are generally analyzed by using the SILAC approach. The SILAC is also applied for secretome comparison of different cells such as pancreatic cells, carcinomas, malignant glioblastomas, and gastric epithelial cells.

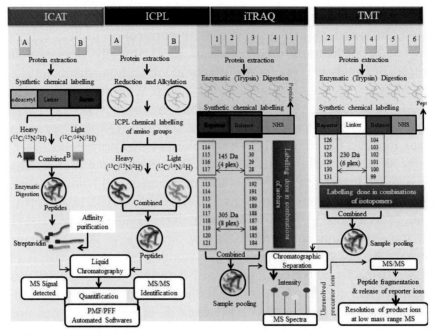

FIGURE 6.5

Comparative workflow of ICAT, ICPL, iTRAQ, and TMT for characterization and practical studies of proteomes.

The differentially expressed proteins in organelles include nucleolus and nucleus.

9.5 Characterization of proteins on the basis of posttranslational modifications

Changes in PTMs play a critical role in the regulatory functions of proteins and changes in cellular dynamics. The characterization of PTMs is subtle to understand the cellular mechanics. MS in combination with SILAC leads to accurate quantifications of the phosphorylation of proteins (Olsen et al., 2006).

9.6 Studying cancer cell proteomes

The technique is useful in studying the quantitative proteomics in cancer cells to trace abnormal signaling cascades.

10. Discussion

To unravel the fundamental processes involved in the normal well-being of living systems, scientists have been exploring a wide range of high-throughput techniques. Huge advancements have been accomplished in omics approaches such as genomics, transcriptomics, connectomics, proteomics, metabolomics, phenomics, and ionomics. This chapter is aimed to understand the technical know-how of proteomics techniques and their application in under pining the phenomenon connected to the physiology and metabolism of fundamental processes. The chapter discussed the basic principle and workflow of gel-based and gel-free proteomic approaches to enhance our understanding of proteomes. These methods play a pivotal role in the field of medicine and the diagnosis of disease. For instance, the detection of novel protein markers and their changing dynamics in the normal and abnormal physiological state helps to resolve several disease ailments such as cancers and their ailments. The metabolic regulations of cells are elucidated by investigating the proteomes of organelles under different developmental stages. The intervention of proteomic technologies augments the field of research pertaining to disease and the normal state of living systems in addition to produce resilient microbes, animals, and plants of desired characteristics.

References

Ashburner, M., Ball, C. A., Blake, J. A., Botstein, D., Cherry, J. M., Davis, A. P., Dolinski, K., Dwight, S. S., Eppig, J. T., Harris, M. A., Hill, D. P., Issel-Tarver, L., Kasarskis, A., Lewis, S., Matese, J. C., Richardson, J. E., Ringwald, M., Rubin, G. M., & Sherlock, G., et al. (2000). Gene ontology: Tool for the unification of biology. The gene ontology consortium. *Nature Genetics, 25*, 25–29. https://doi.org/10.1038/75556, 10802651.

Bottari, P., Aebersold, R., Turecek, F., & Gelb, M. H. (2004). Design and synthesis of visible isotope-coded affinity tags for the absolute quantification of specific proteins in complex mixtures. *Bioconjugate Chemistry, 15*, 380–388.

Colangelo, Christopher M., & Williams, Kenneth R. (2006). "Isotope-coded affinity tags for protein quantification". New and emerging proteomic techniques. *Methods in Molecular Biology, 328*, 151–158. https://doi.org/10.1385/1-59745-026-X:151, 978-1-59745-026-3.

Cox, J., & Mann, M. (2008). MaxQuant enables high peptide identification rates, individualized p.p.b.-range mass accuracies and proteome-wide protein quantification. *Nature Biotechnology, 26*, 1367–1372.

Cui, Z., Chen, X., Lu, B., Park, S. K., Xu, T., Xie, Z., Xue, P., Hou, J., Hang, H., Yates, J. R., & Yang, F., et al. (2009). Preliminary quantitative profile of differential protein expression between rat L6 myoblasts and myotubes by stable isotope labelling with amino acids in cell culture. *Proteomics, 9*, 1274–1292. https://doi.org/10.1002/pmic.200800354, 19253283.

Dayon, L., Hainard, A., Licker, V., Turck, N., Kuhn, K., Hochstrasser, D. F., Burkhard, P. R., & Sanchez, J. C. (2008). Relative quantification of proteins in human cerebrospinal fluids by MS/MS using 6-plex isobaric tags. *Analytical Chemistry, 80*(8), 2921–2931.

Gehrmann, M. L., Hathout, Y., & Fenselau, C. (2004). Evaluation of metabolic labeling for comparative proteomics in breast cancer cells. *Journal of Proteome Research, 3*, 1063–1068.

Gygi, S. P., Rist, B., Gerber, S. A., Turecek, F., Gelb, M. H., Aebersold, R., & October. (1999). Quantitative analysis of complex protein mixtures using isotope-coded affinity tags. *Nature Biotechnology, 17*(10), 994–999. https://doi.org/10.1038/13690

Huang da, W., Sherman, B. T., & Lempicki, R. A. (2009). Systematic and integrative analysis of large gene lists using DAVID bioinformatics resources. *Nature Protocols, 4*, 44–57.

Jensen, L. J., Kuhn, M., Stark, M., Chaffron, S., Creevey, C., Muller, J., Doerks, T., Julien, P, Roth, A., Simonovic, M., Bork, P., von Mering, C., et al. (2009). String 8—A global view on proteins and their functional interactions in 630 organisms. *Nucleic Acids Research, 37*, D412–D416. https://doi.org/10.1093/nar/gkn760, 18940858.

Kanehisa, M., Goto, S., Kawashima, S., Okuno, Y., & Hattori, M. (2004). The KEGG resource for deciphering the genome. *Nucleic Acids Research, 32*, D277–D280.

Keller, A., & Shteynberg, D. (2010). Software pipeline and data analysis for MS/MS proteomics: The trans-proteomic pipeline. *Methods in Molecular Biology, 694*, 169–189.

Kelstrup, C. D., Young, C., Lavallee, R., Nielsen, M. L., & Olsen, J. V. (2012). Optimized fast and sensitive acquisition methods for shotgun proteomics on a quadrupole orbitrap mass spectrometer. *Journal of Proteome Research, 11*, 3487–3497.

Klose, J. (1975). Protein mapping by combined isoelectric focusing and electrophoresis of mouse tissues. *Human Genetics, 26*(3), 231–243. https://doi.org/10.1007/BF00281458

Leroy, B., Rosier, C., Erculisse, V., Leys, N., Mergeay, M., & Wattiez, R. (2010). Differential proteomic analysis using isotope-coded protein-labeling strategies: Comparison, improve- ments and application to simulated microgravity effect on *Cupriavidus metallidurans* CH34. *Proteomics, 10*(12), 2281–2291.

Liu, C., Song, C. Q., Yuan, Z. F., Fu, Y., Chi, H., Wang, L. H., Fan, S. B., Zhang, K., Zeng, W. F., He, S. M., & Dong, M. Q., et al. (2014). pQuant improves quantitation by keeping out interfering signals and evaluating the accuracy of calculated ratios. *Analytical Chemistry, 86*, 5286–5294.

Lossner, C., Warnken, U., Pscherer, A., & Schnolzer, M. (2011). Preventing arginine-to-proline conversion in a cell-line independent manner during cell cultivation under stable isotope labeling by amino acids in cell culture (SILAC) conditions. *Analytical Biochemistry, 412*, 123–125.

Lottspeich, F., & Kellermann, J. (2011). ICPL labeling strategies for proteome research. *Methods in Molecular Biology, 753*, 55–64.

Magdeldin, S., Zhang, Y., Bo, X., Yoshida, Y., & Yamamoto, T. (2012). *Biochemistry, genetics and molecular biology "gel electrophoresis— Principles and basics"*. Rijeka, Croatia: INTECH.

Mann, M., Kulak, N. A., Nagaraj, N., & Cox, J. (2013). The coming age of complete, accurate, and ubiquitous proteomes. *Molecular Cell, 49*, 583–590.

Michalski, A., Damoc, E., Hauschild, J. P., Lange, O., Wieghaus, A., Makarov, A., Nagaraj, N., Cox, J., Mann, M., & Horning, S., et al. (2011). Mass spectrometry-based proteomics using Q Exactive, a high-performance benchtop quadrupole Orbitrap mass spectrometer. *Molecular and Cellular Proteomics, 10*, M111, 011015.

O'Farrell, P. H. (1975). High resolution two-dimensional electrophoresis of proteins. *Journal of Biological Chemistry, 250*(10), 4007–4021. PMID: 236308.

Olsen, J. V., Blagoev, B., Gnad, F., Macek, B., et al. (2006). Global, in vivo, and site-specific phosphorylation dynamics in signaling networks. *Cell, 127*, 635–648.

Ong, S. E., Blagoev, B., Kratchmarova, I., Kristensen, D. B., Steen, H., Pandey, A., & Mann, M. (2002). Stable isotope labeling by amino acids in cell culture, SILAC, as a simple and accurate approach to expression proteomics. *Molecular and Cellular Proteomics, 1*(5), 376–386.

Ong, S. E., Mittler, G., & Mann, M. (2004). Identifying and quantifying in vivo methylation sites by heavy methyl SILAC. *Nature Methods, 1*, 119–126.

Park, S. S., Wu, W. W., Zhou, Y., Shen, R. F., et al. (2012). Effective correction of experimental errors in quantitative proteomics using stable isotope labelling by amino acids in cell culture (SILAC). *Journal of Proteomics, 75*, 3720–3732.

Qiu, Y., Sousa, E. A., Hewick, R. M., & Wang, J. H. (2002). Acid-labile isotope-coded extractants: A class of reagents for quantitative mass spectrometric analysis of complex protein mixtures. *Analytical Chemistry, 74*, 4969–4979.

Schmidt, A., Kellermann, J., & Lottspeich, F. (2005). A novel strategy for quantitative proteomics using isotope-coded protein labels. *Proteomics, 5*(1), 4–15.

Shannon, P., Markiel, A., Ozier, O., Baliga, N. S., Wang, J. T., Ramage, D., Amin, N., Schwikowski, B., & Ideker, T., et al. (2003). Cytoscape: A software environment for integrated models of biomolecular interaction networks. *Genome Research, 13*, 2498–2504. https://doi.org/10.1101/gr.1239303

Tyanova, S., Mann, M., & Cox, J. (2014). MaxQuant for in-depth analysis of large SILAC datasets. *Methods in Molecular Biology, 1188*, 351–364.

Analysis of proteomes—II

1. Introduction

All the spectroscopic techniques employ light to interact with the molecule of interest to investigate its structure or other parameters such as concentration. The molecular features of a compound or other molecules are elucidated by using electromagnetic radiations, which have different energies that are used to analyze the diverse features of compounds. Consequently, the class/type of spectroscopic technique grossly lies in the type of radiation used to elucidate the molecular information of molecules. For instance, the UV or visible light is used to probe the conformational structure of biomolecules. So it is critical to understand the properties of electromagnetic radiation that are sued for investigating the matter under investigation. The quantum phenomenon, properties of radiation, and structural parts of the sample are based on the interaction of electromagnetic radiation under operation. Electromagnetic radiation is composed of magnetic and electric vectors oscillating perpendicular to each other. The wavelength of a particular type of wave is measured as the distance between two consecutive peaks and is usually calculated in nanometers (nm) and designated as "λ." The maximum length of the vector is known as amplitude. The frequency (ν) of the electromagnetic radiation is calculated as a number of oscillations done in the time frame of 1 s.

2. Ultraviolet and visible light spectroscopy

This range of the electromagnetic spectrum is widely used for analytical work to investigate biological samples. Among four possible transitions $n \rightarrow \pi$, $\pi \rightarrow \pi$, $n \rightarrow \sigma$, $\sigma \rightarrow \sigma$, only $n \rightarrow \pi$, $\pi \rightarrow \pi$ transitions are

Principles of Genomics and Proteomics. https://doi.org/10.1016/B978-0-323-99045-5.00010-0

elicited by ultraviolet and visible light. It is actually the chromophores, actually substructures that interact with the electromagnetic radiations. For instance, in context with proteins, three types of chromophores, viz., peptide bonds, side chains of aromatic amino acids, prosthetic groups, and coenzymes, exist for UV/Vis spectroscopy. The presence of conjugated double bonds in organic molecules results in extended π systems.

The peptide bond electronic transitions occur in far UV radiations. The intense peak in proteins occurs at 190 nm and weaker ones at 210–220 nm grossly due to $n \rightarrow \pi$ and $\pi \rightarrow \pi$, which exist in amino acids such as Arg, Asp, Asn, Gln, Glu, and His. These transitions are not observed in proteins because they are masked by highly intense absorption by peptide bonds. The absorption spectrum of aromatic amino acids like phenylalanine occurs at 257 nm, tyrosine at 274 nm, and tryptophan at 280 nm. Even though prosthetic groups such as flavin, haem, and metal protein complex carotenoid possess strong absorption maximum at UV/Vis spectra. These bands are usually sensitive to the local environment and can be used for physical studies of enzyme action. Carotenoids, for instance, are a large class of red, yellow, and orange plant pigments composed of long carbon chains with many conjugated double bonds. They contain three maxima in the visible region of the electromagnetic spectrum (420 nm, 450 nm, 480 nm).

2.1 Basic principle

The principle of UV–visible spectroscopy is based on UV light absorption or visible light by sample, which in turn leads to distinct spectra. Consequently, the spectra are dependent on the interaction of light and matter. Upon this process, the sample absorbs light and undergoes excitation, that is, the electron moves from the ground state to an excited state and then falls again to the ground state resulting in the production of the spectrum. It is critical to remember that energies of the ground state and excited state are equal to the amount of visible or UV radiation absorbed by the sample. The chance of absorbing a photon is given by the extinction coefficient, which is dependent on the "λ" of a photon. For instance, if light with an intensity of I_0 passes through a sample with appropriate transparency and thickness of cuvette, that is, path length "d." The absorption characteristic parameter of the sample is designated by extinction coefficient "α." The concentration of the sample is given at a molar concentration "c." The phenomenon of absorption and production of spectra is given by Beer—Lambert law, which states that when a monochromatic light passes incident through a sample solution that absorbs light, the concentration of absorbing substance in the sample solution is directly proportional to the intensity of monochromatic light. Consequently, the higher the number of absorbing molecules in the sample higher is the extent of absorption of light by samples. It must be noted that Beer—Lamberts law is valid for only low concentred samples. The reason behind this limitation of Beer—Lambert law lies in the association of molecules in highly concentrated samples causing deviation from the ideal behavior of samples. The law is linearly proportional to the chromophores concentrations in the compound. According to the Beer—Lambert law, absorbance is linearly proportional to the concentration of chromophores.

The absorption or light scattering is usually designated in form of optical density. One must remember that in some cases the measurement of turbidity of cells in cultures is not usually dependent on light absorption but on light scattering when detected by the spectrophotometer. In highly turbid samples, the bacterial cells do not absorb the incoming light, rather light is scattered and hence the spectrophotometer records the apparent absorbance of the cell cultures. The recorded values or parameter is known as optical density (Fig. 7.1).

2.2 Major factors that affect the absorption by UV/Vis spectrum

It is important that the samples be dissolved in a water solvent, as the water spectrum does not show any absorption bands and thus acts as an inert medium. In addition, the spectrum of a particular chromophore is shown by the chemical structure of the compound. Other factors that influence the spectrum of the sample include solvent polarity, orientation effects, and protonation/deprotonation. Thus consequently the chromophore probing can be assessed by their absorption. The effects that are considered to affect the chromophore and its interaction with

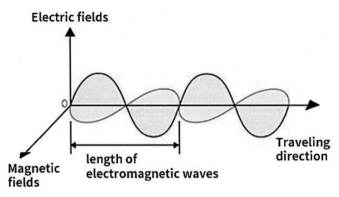

FIGURE 7.1

Electromagnetic radiation showing magnetic and electric vectors oscillating perpendicular to each other.

wavelength are the shift of wavelength to higher values also known as a bathochromic shift of redshift, shift to lower wavelength (hypsochromic effect), and increase in absorption (hyperchromic shift). In addition, an increase in the order of DNA/RNA single strands to double-stranded forms also leads to differential absorption behaviors. Since dsDNA has smaller absorption power than ssDNA strands. Solvent polarity is another parameter that effects both the ground and excited states of the compound under investigation (Fig. 7.2).

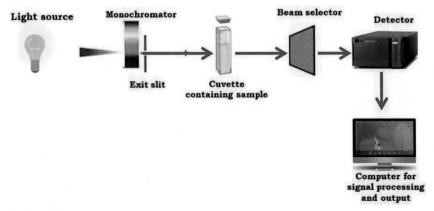

FIGURE 7.2

Workflow and components in a UV—Vis spectrophotometer.

2.3 **Basic workflow**

The spectrophotometers use UV or visible light sources to illuminate the sample under investigation. The instrument subsequently measures the light reflected, absorbed, or transmitted. The light sources commonly used include deuterium arc lamps, xenon flash lamps, and tungsten-halogen lamps. The deuterium uses deuterium gas and yields UV rays and radiations in the visible region. The light intensity of deuterium arc lamps steadily decreases over time, so D2 lamps need to be replaced frequently. A tungsten-halogen lamp employs a filament. Once electricity passes through the filament, it emits light due to heating up. This type of lamp is usually used in combination with D2 lamps. Xenon flash lamp is another type of light source used in a spectrophotometer, which emits light for a very short time period and has proven to possess a very long life and needs very infrequent replacements (Fig. 7.3).

2.4 **The monochromator**

In a spectrophotometer, only one selected wavelength band is used for sample analysis. For this objective, the monochromators are used to select a particular wavelength. The monochromator consists of the entrance slit, a dispersion device for spreading light in different wavelengths that displays a rainbow of light and it also allows the selection of a specific band of light. The monochromator also consists of an exit

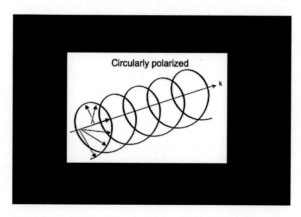

FIGURE 7.3

Circularized polarized light.

slit, from which the light of a specific wavelength passes and is targeted onto the sample. The sample is positioned in such a way that a beam of monochromatic light will pass through it without any interface. For the liquid samples, the sample is held in the cuvette with a fixed path length. The cuvette is made of materials such as quartz, glass, and plastic to allow the transmission of visible or UV light. The existing standard cuvettes have a path length of 10 mm to allow maximum needed transmittance (Fig. 7.4).

2.5 Basics for analysis of samples

It is critical to measure the absorbance of the reference sample usually called as a blank sample. The blank sample cuvette is filled with a similar solvent in which the desired sample is dissolved. The approach obviates any variations in the light intensity because both the sample and reference will be affected equally.

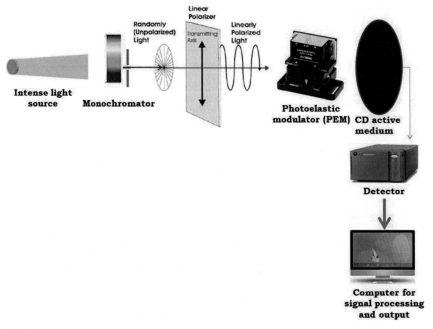

FIGURE 7.4

Representation of workflow of common CD spectrometer showing the polarization of light and the differential absorption of light.

2.6 **Some strengths/limitations**

The technique does not destruct the sample, such that the sample may be reused. Measurements can be analyzed easily and quickly. Minimal processing of data is needed, hence less training to operate spectrophotometers.

2.7 **Applications of UV–visible spectrophotometer**

1. **The identification of structure and spectra:** it provides a limited amount of qualitative information and few broad absorbance peaks. Most of the compound absorption is due to π bonds.
2. **Partial confirmation of identifying compounds:** It is accomplished by comparing the measured spectrum of the compound with that of the reference molecule. The yielded complex spectra are useful for the characterization of materials or for identifying the compounds under investigation.

2.8 **Quantification and purification of DNA and RNA analysis**

For downstream processing of nucleic acids, it is critical that samples must be identified and purified for high throughput results. The purity of DNA/RNA is obtained by checking absorbance at 260 nm/280 nm. If the ratio indicates 1.8, the reading confirms the presence of pure DNA, and for pure RNA sample, it is 2.0. The absorbance of DNA is a little less than RNA due to the presence of uracil in RNA and its absence in DNA. Proteins absorb less at 260 nm and have a lesser ration or below 1.8 (Table 7.1).

Table 7.1 Details of parameters used to assess the nucleic acids or proteins in spectrophotometric analysis.

Wavelength (nm)	Structures mainly showing UV absorbance	Biomolecule present
230	Shape of protein	Protein
260	Nucleotides such as uracil, adenine, guanine, cytosine, and thymine	DNA and RNA
280	Aromatic amino acids such as tryptophan and tyrosine	Protein

2.9 **Analysis of pharmaceuticals**

By overlapping the peaks spectra obtained from UV−Vis absorbance spectrophotometry of drugs with reference molecules, the identification may be carried.

2.10 **Analysis of beverages**

For instance, the concentration of addictive compounds such as caffeine in soft drinks can be analyzed by UV−visible spectroscopy. Other compounds identified such as anthocyanins are used for quality control of various plant products.

2.11 **Checking the number or concentration of bacterial cells in cultures**

Optical density at 600n radiation is widely sued to calculate the cell density and hence track the growth kinetics. The estimation is user-friendly and very less quantity is used during analysis.

2.12 **Applications**

Ultraviolet−visible (UV−Vis) spectroscopy is a widely used technique in many areas of science ranging from bacterial culturing, drug identification, and nucleic acid purity checks and quantitation, to quality control in the beverage industry and chemical research. This article will describe how UV−Vis spectroscopy works, how to analyze the output data, the technique's strengths and limitations, and some of its applications.

3. **Fluorescence spectroscopy**
3.1 **Basic principle**

We have noticed in many chemical properties of atoms and molecules that can undergo both radiative decay and thermal decay after the electrons are excited by UV or visible light. The molecules have the capability to emit photons with longer wavelengths than the exciting light. In this state, they can fluoresce in the UV−Vis or near infra-red region of the spectrum. This phenomenon is greatly exploited by various techniques to identify or quantify the biomolecules in living systems under laboratory conditions. Fluorescence spectroscopy provides the information about nature of the excited state and electronic excitation energy of a molecule.

This technique is very fast, user-friendly, and more importantly inexpensive that is employed for determining analyte concentration based on the fluorescing properties of the compound in solution. In this technique, a beam of radiation with a wavelength measuring between 180 and 800 nm is used to pass through the analyte in a cuvette. The instrument then measures from an angle the light that is emitted by the analyte. Both the excitation spectrum and emission spectrum are employed in this method. Hence, the concentration of the analyte is directly proportional to the intensity of the emission. The analysis of the compound grossly depends on the emission of spectra, which in turn depends on the analyte concentration, the wavelength used for excitation, path length, and absorption of radiation by the sample.

3.2 Basic workflow for fluorescence detection and its measurement

3.2.1 The use of filter fluorometer: following parts are found in this part of fluorescent spectrophotometer

1. Excitation source such as a lamp or some laser. In modern fluorescence spectrophotometers, high-pressure xenon arc lamps are used.
2. Primary filter permits radiation having a specific wavelength and eliminates the residual scatters and nowadays monochromators are used (records both excitation and emission spectra) (Lakowicz, 2013).
3. Sample chamber or cuvette is used to hold the analyte.
4. Fluorescence detector—it consists of various photomultiplier tubes that aid in the amplification of photon emission, and record and display the electronic signals (Albani, 2007; Sharma & Schulman, 1999).

The fluorescent measurements can be steady state or time resolved, in the case of steady state, the sample is illuminated with a continuous beam of light and the emission spectra by the sample is recorded for analysis. Whereas, in time-resolved measurement, the analyte is illuminated by a very short pulse of radiation after which the anisotropy or intensity decay is measured.

3.3 Major applications of fluorescence spectroscopy

The technique is widely used in medical microbiology as a diagnostic tool due to its high sensitivity and specificity for detecting microorganisms. In combination with fluorescence correlation spectroscopy may aid in understanding the pathophysiological steps of a wide range of microorganisms (Alexandra et al., 2004, 2005).

1. Quantitation of nucleic acids:
   It Aids in high-precision quantification of DNA and RNA by using fluorophores such as ethidium bromide is added to samples such as DNA/RNA and subsequently, the samples are loaded into the spectrophotometer. In addition, advanced fluorescence spectroscopic techniques are employed in single molecule real-time analysis of DNA. The technique is central to the genetic diagnostic revolution (Ardui et al., 2018).
2. Contamination of industrial products:
   The technique is sued for a fast, noninvasive method for the quality assessment of contaminants (Kohli, 2012). For instance, the technique is used for the detection of contaminants such as organic compounds in groundwater (Dahm, Van Straaten, Munakata-Marr, & Drewes, 2012).
3. Application in the field of medicine:
   The method is sued to study nanoparticle syntheses that are used for drug delivery and other medical applications. It is critical to study the interaction of protein corona and nanoparticles for safe and efficient use (Röcker et al., 2009).
4. Environmental monitoring
   The 3D-excitation emission matrix fluorescence spectroscopy and high-resolution fluorescence spectroscopy are employed to analyze the dissolved organic matter and hence it also helps in optimizing the treatment process of landfill chelates (Leenheer & Croué, 2003).

5. Analysis of pharmaceutical drugs:
   The method is sued to analyze the drugs, for instance, the use in analyzing the coformulated tablets, generally employed for cholesterol medication. It is a very quick and reliable method for the quality control of medicines (Ayad & Magdy, 2015).
6. Agricultural systems:
   The method is used for the identification of diverse crop varieties. For instance, the laser-induced fluorescent emission technique is the best technique to identify the varieties of citrus by analyzing seedlings (Milori et al., 2013). Similarly, total luminescence spectroscopy has been employed to characterize the tea varieties by manufacturers.

3.4 Utility of fluorescence spectroscopy

- Analyte can be dissolved in a wide range of solvents such as ethanol, water, and hexane
- One can use only UV or visible light for analysis.
- The emission usually emits in visible IR radiation
- The quantitative measurements of a single analyte in solution

4. Nuclear magnetic resonance spectroscopy

The technique has evolved as a prominent technique for the determination of the structure of organic molecules. The technique helps in the complete analysis as well as interpretation of the complete spectrum. Even though a higher amount of samples are required than other mass spectroscopy techniques, nuclear magnetic resonance spectroscopy (NMR) processing does not destruct the samples. The method is based on several physical principles. For instance, many elemental nuclei have a characteristic spin (I). Some of the nuclei have fractional spin such as $I = 1/2, 3/2, 5/2 \ldots$, while others have integral spin such as $I = 1, 2, 3, \ldots$ and few nuclei have no spin at all $I = 0$. The following section will explain the significance of studying NMR with respect to isotopes having $I = 1/2$. The strong magnetic fields are very much critical for the operation of NMR spectroscopy. Some critical points to consider in NMR spectroscopy:

- The spinning charge generates a magnetic field, and the resulting spin magnet has a magnetic moment proportional to the spin of the nuclei.
- In presence of an external magnetic field, at least two spins exist, viz., $+1/2$ and $-1/2$. During the process, the magnetic moment of the $+1/2$ state (lower energy state) aligns with the external field.
- The energy difference between the two spins is largely dependent on the external applied magnetic field and the observed difference is very small. The difference in the two energy states is increased once the divergence of the field increases.

4.1 The basic principle

The frequent studies in organic chemistry largely involve 1H, but most NMR spectroscopic studies involve 13C, 15N, and 31P isotopes. The resonance condition of the compound is usually satisfied by the external magnetic field, with absorption occurring at a frequency of 40 MHz (region of radio waves). Electromagnetic radiation promotes the transition from a lower state to a higher energy state. After a certain time span, the spin from the higher state is again lowered to low energy, this state is known as relaxation. The energy released during this state is emitted as heat and is known as spin-lattice relaxation. The process occurs with the rate of T11, and T1 known as longitudinal relaxation time. Magnetism also changes with time, largely due to the interaction of several nuclei. The latter phenomenon is generally called as spin—spin relaxation and is an inert characteristic of transverse relaxation time T2. The molecular environment created by protons governs the value for externally applied filed resulting in the resonating nucleus. This is recorded in the form of a chemical shift (d). The chemical shift arises from the applied field inducing secondary fields of about 0.15—0.2 mT at the proton by interacting with the adjacent bonding electrons. If in case the induced field opposes the externally applied field, the applied field needs to be of higher value for resonance to occur. In this condition, the nucleus is said to be shielded and the magnitude of shielding is proportional to the electron-withdrawing power of proximal substituents. In case, the applied and induced fields are aligned, the applied field needs to be of lower value for resonance to occur. In this condition, the nucleus is said to be shielded.

4.2 About magnetic field

Tesla (T) is the international unit for magnetic flux. The magnetic field of the earth is not constant and ranges from 10^{-4} at ground level. Currently, existing NMR spectrometers use highly powerful magnetic fields ranging from 1 to 20 T; the energy difference is still less than 0.1 cal/mol. The small energy differences (ΔE) are usually given in frequency units of MHz and range from 20 to 900 Mz, which grossly depend on strength of the magnetic field and the nucleus under investigation. In addition, irradiation of any sample with radio frequency (rf) exactly corresponds to the spin state of separation of a specific set of nuclei.

For instance, the nucleus of a hydrogen atom possesses the magnetic moment $\mu = 2.7927$ is most often studied by NMR spectroscopic method. The energy differences between two spin states of the magnetic field of spin ½ nuclei in a given magnetic field are usually proportional to magnetic moments. The magnetic moments of a few common nuclei are given below:

- ^{1}H $\mu = 2.7927$
- ^{19}F $\mu = 2.6273$
- ^{31}P $\mu = 1.1305$
- ^{13}C $\mu = 0.7022$

4.3 NMR spectroscopy of proton

The NMR spectrometer must be tuned to proton, and the protocols for obtaining the actual spectrum vary. The continuous wave method is used to obtain the varying spectrum of any nucleus. The protocol involves using a sample solution in a uniform 5 mm glass tube oriented between poles of string and a powerful magnet. The magnet is spun at an average magnetic field variation. The antenna coil is used to direct the radiation frequency of appropriate energy into the sample under investigation. The sample tube is covered with a receiver coil and the emission of absorbed rf energy, which is monitored by electronic devices and a connected computer. The spectrum is obtained by sweeping or varying the magnetic field over a small range while observing the rf signals obtained from the samples.

4.4 Case study—1: H₂O molecule

For instance, sample water was irradiated with 100 MHz under the influence of a 2.3487 T external magnetic field. If the slight increase in the magnetic field of 2.3488 T is increased, water molecules will absorb rf energy and show resonance signals. All the protons have the same magnetic moment; we might expect all the hydrogen atoms to display a similar resonance signal, which is not real in this scenario. By obtaining independent measurements and calculations, it is determined that the naked protons resonate at low field strength energy than a covalently linked hydrogen molecule. Almost all the compounds are measured in form of gases except, water, sulfuric acid, and chloroform.

The proton nuclei in different compounds behave differently due to the surrounding of electrons around the protons in covalent compounds and ions. Electrons move in response to an external magnetic field to generate a secondary field that opposes a much stronger field. The generated secondary field shields the nucleus from the applied field to increase the resonance. Hence magnetic field must be increased to compensate for the induced magnetic shield. The compounds CH4, HCl, HBr, and HI give resonance signals at the higher field and have proton nuclei more shielded than other compounds. Most of the organic compounds show proton resonance ranging within 12 ppm, and hence it is very critical to measure these compounds with very precision considering their sensitivity.

4.5 The concept of chemical shift

The location of a wide range of NMR resonance signals is largely dependent on both the external magnetic field and radio frequency (rf) unlike the other spectroscopic techniques such as infrared and UV−visible spectroscopy, where absorption peaks are uniquely located by a frequency or wavelength. No two magnets have the same fields, so resonance frequencies vary accordingly means there is a need of specifying and characterizing the locations of NMR signals. The problem is resolved by reporting the NMR signal in a spectrum relative to the reference signal from a standard compound added to the sample under investigation. The reference compound should be chemically unreactive and can be easily removed after the measurement of the sample. In addition, the reference compound must provide a single and sharp single signal that must not be interfering with the resonance of the compound under investigation. Compounds such as tetramethylsilane, $(CH_3)_4Si$, meet almost the characteristic of a good reference compound for carbon and proton NMR peaks. Chemical shifts are given in parts-per-million (ppm) and are designated by the δ symbol.

The NMR spectra of the solid compound are done by dissolving it in a suitable solvent. The dissolving usually used includes carbon tetra chloride (CCl_4). The feasibility of this compound lies in absence of hydrogen ions. Other solvents used include deuterium-labeled compounds, such as chloroform-d ($DCCl_3$), benzene-d_6 (C_6D_6), deuterium oxide (D_2O), DMSO-d_6 (CD_3SOCD_3), and acetone-d_6 (CD_3COCD_3). As the deuterium isotope of hydrogen has a different magnetic moment and spin, it is invisible in a spectrometer tuned to protons.

4.6 The signal strength

The NMR resonance with respect to intensity and magnitude is displayed in the vertical axis of a spectrum and is largely [proportional to the sample concentration. Thus, a small change in dilutions will result in a change in signal peaks. For instance, the equal concentration of cyclohexane and benzene in CCl_4 solution shows resonance signals from cyclohexane at twice the rate as compared to benzene. This is because cyclohexane has twice hydrogen per molecule as compared to benzene.

4.7 The π-electron functions in NMR

In certain conditions, the low field resonance of hydrogen bonded to aromatic ring carbons is confusing to resolve. The anomalies associated with these conditions involve the formation of pi-electron systems with hydrogen. These conditions are due to the interaction of pi-electrons with external magnetic fields.

4.8 The significance of NMR

The NMR technique is of great value to elucidate chemical structures. By using NMR, both qualitative and quantitative information of compounds may be obtained. Advanced forms of NMR have enhanced the accuracy and efficiency in the analysis of compounds. Scanning and accumulation of data can aid in enhancing the weak signals; the process is called as computer averaging of transients or CAT scanning and helps to improve the signal-to-noise ratio.

4.9 Applications of NMR

Determination of molecular structure: it is mainly used for determining the structure of organic compounds. The chemical shifts generated during the process aid in evidence of protons or carbons resulting in the elucidation of functional groups. The proton linking with the carbon skeleton is known by spin—spin interactions. The intensity of signals provides the background for calculating the number of protons participating in the generation of a particular signal.

Structure of proteins and peptides in solution: NMR can be sued to elucidate the structures of proteins up to an MW of about 50 kDa. The magnet size with a field strength of up to 900 MHz directly enhances the capability of NMR to elucidate the higher MW proteins. The required amount of samples can be reduced by the introduction of cryotube technology just to avoid the use of large amounts of samples, viz., 10 mg for NMR analysis.

4.10 NMR for elucidating the metabolite structures

This technique is used to determine the conformation and chemical structure of molecules. This technique is the next leading technique after X-ray crystallography used to determine absolute stereochemistry.

5. X-ray diffraction

The DNA structure was primarily described by X-ray crystallographic data obtained by the X-ray diffraction (XRD) method. The data are used to determine the structure of proteins and other biological samples. The data aid in deciphering the 3D structure of molecules. In this technique, the purified samples are firstly crystallized and subsequently, the crystallized samples are exposed to X-ray beams. In this method, a wavefront of X-rays is approached as a series of similar equidistant and parallel lattice planes (Bunaciu et al., 2015). Part of this beam will be reflected, at a specified angle of reflection. The diffraction patterns obtained from the sample are processed to yield the crystal packing symmetry and repeat units to form a complete crystal. The data are obtained in the form of diffraction spots. The intensity of these spots is used to generate the structure factors, which are used to map the electron density calculation. For obtaining sufficient clarity to build a molecular structure, the protein structure data are used. The resulting structure is further refined to fit in the map accurately to yield thermodynamically favored conformation of the said compound.

5.1 The crystallization of proteins

A good protein source must be available with the help of purification and concentration methods for obtaining high-quality soluble protein. One must take highly concentrated samples, such that the sample comes out of solution. Several factors are critical to consider when subjecting a sample to crystallization such as

- Choice of precipitant
- Concentration of precipitant
- Buffer
- The pH
- Concentration of protein
- Temperature conditions
- Crystallization technique
- Inclusion of additives

Various forms of the molecule may exist such as no spots, precipitation, very tiny crystals, and a wide array of microcrystals. The crystal size can be obtained by various techniques such as

- Seeding
- Temperature alteration
- Change in concentration of proteins

It must be kept in consideration that for obtaining diffraction analysis the crystallized proteins must be 0.1 mm long to provide a sufficient amount of crystal lattice when exposed to an X-ray beam. So it must be verified that crystals contained our desired macromolecule. This may be ensured by sacrificing a few crystals to run in PAGE or staining or again test in X-ray diffraction.

5.2 The optical setup in X-ray crystallography

The generation of X-rays occurs from an accelerating electron from electrons striking a copper anode or in a synchrotron storage ring. In the synchrotron storage ring, a single X-ray is selected by the mechanism of absorption of unwanted X-rays by the process called as monochromation. In the case of electrons striking a copper anode, only one of the predominant X-rays is produced. In both cases, the motive is that a beam of X-ray is focused and collimated for obtaining a parallel beam of X-rays by the aid of slits having a diameter of 0.1−0.3 mm; the crystal is then mounted in the parallel electron beam and finally adjusted carefully on goniometer head. In addition, a backstop, made of lead pellets, is placed between the crystal and X-ray detector to prevent an intense source from reaching the detector.

5.3 The mechanism of diffraction analysis

Once we have confirmed the formation of an appropriate crystal of suitable size of some macromolecule, we need to go for X-ray diffraction analysis. The exposure of new crystals can be performed by using trail exposures in synchrotron sources or an X-ray generator. In another approach, the crystals are mounted in a capillary tube at laboratory temperature or one can mount them frozen in a small loop under the influence of a stream of liquid nitrogen at 100 K. The images are by imaging plate, which has proven to be more sensitive than X-ray films and the images are obtained within a few minutes. Currently, charged coupled device (CCD) technology-based detectors are sued to obtain the data (Gruner, 1994; Moy, 1994). The distance between the crystal and the detector is adjusted to a maximum of 1.5−3.0 Å resolution. As the diffraction angle increase, the resolution of spots also increases on the detector. It is at the edge of the detector where the highest resolution of spots is observed. Many obtained image display programs incorporate an algorithm for the resolution of each spot. In general, spots beyond 3 Å are required: a carbon–carbon bond is approximately 1.5 Å, but a resolution of close to 3 Å is sufficient to be able to detect the amino acid side chains in the electron density map. At high resolution, the diffraction image becomes weaker, which is limited based on how molecular subunits are ordered. The determination of crystal system, space group, and unit cell dimension. The smallest unit that forms a crystal lattice is a unit cell. The dimensions of which are given in 03 lengths, viz., a, b, c with 03 angles α, β, and γ. About 230 space groups exist, even though all these are not permitted for proteins, largely due to the chirality of amino acid residues.

5.4 Crystal structure determination

The Braggs equation is used for measuring the maximum reflection from different angles. The distance between lattice planes of a particular crystal is calculated by employing two methods including the powder method and the rotating crystal method.

5.5 The rotating crystal method

In this approach, the monochromatic X-radiation is incident on the crystal and is rotated on its axis such that it is exposed to X-rays in all dimensions. The beams reflected from the source lie as spots that are almost coaxial to the rotation axis. To determine positions of maximum intensity due to reflection one can use a photographic film or Bragg X-ray spectrometer. The method uses large crystals having well-defined faces. As reflections of several orders must be examined for a large number of faces, the total labor involved is enormous. This, however, is compensated by a relatively simple interpretation of the results.

In the powder method, the X-ray crystallography method is widely used, in which a narrow beam of X-rays is directed at the finely powdered substance. This powder is usually coated on thin walls of glass tube or hair. The X-rays diffracted from the sample substance collide with photographic film.

5.6 Retrieving and processing of data

The retrieving of data from X-ray diffraction is dependent on the following conditions:

- **Crystallographic symmetry:**
  The space groups and the amount of symmetry in the crystal systems are one of the major conditions that define the data of the molecule. In the case of high symmetry systems, it is important to collect the diffraction data as little as 35 degrees. While in the case of lower symmetry crystal systems, data need to be collected through 180 degrees.
- **Noncrystallographic symmetry** (Smyth et al., 1995)**:**
  A sample or particle such as a virus is composed of many identical subunits and is having high noncrystallographic symmetry (NCS). In addition, monomeric proteins exhibit no NCS and hence the requirement for the complete data set.

  It is assumed that diffraction data is enhanced exponentially with respect to resolution. In ideal conditions, a complete data set will be retained from a single crystal. The data processing is mathematically complex by using algorithms and software. In data processing, the first step is the determination of crystal systems and the determination of unit cell dimensions accurately. This is the stage in which we determine the crystal orientation in the X-ray beam (Leslie, 1993). The best time to know the indexing is when the cell and orientation are known. Each spot on the image is designated as an index and is quoted with 03 integers' viz., h, k, and l. Thereafter, computer programming is employed for autoindexing of diffraction images to get the real image.

5.7 Applications of X-ray diffraction method

- It is used for the identification of unknown crystalline materials such as inorganic ions, minerals, geological samples, and biological samples
- Aids in the characterization of crystalline materials
- Helps in determining the unit cell dimensions
- In purity of compounds

- Helps in identifying the calcium lactate pentahydrate crystals in Cheddar cheese (Tansman et al., 2014).
- In the pharmaceutical industry, XRD is most successful in establishing the formulations for discovering the degree of crystallinity and helping in the identification of polymorphs. It also aids in determining the quantity of each compound in a mixture. It is also sued to study the influence of moisture on drugs.
- In single-crystal diffraction analysis, the crystal grown in laboratory conditions is mounted on a goniometer and then exposed to the X-rays to obtain a diffraction pattern of regularly spaced spots. These spots are then detected by detectors, and an image is developed on image plates and CCD cameras for compounds, proteins and other small molecules.
- In forensic science, the XRD is sued to identify the drugs in forensic analysis studies.
- In geology, the XRD is a key technique to identify the minerals and characterize the individual crystal structures. Hence, it aids in determining the proportion of different minerals in the soil samples.

6. **Circular dichroism**

Circular dichroism (CD) is a type of absorption spectroscopy that uses circularly polarized light to evaluate the structural features of optically active samples. The technique is mostly sued to investigate the biological samples for obtaining their structure and especially to study their interaction with metal ions and other molecules. The method is based on differential absorption of left and right circularly polarized light that yields a difference in the absorption coefficients $D\varepsilon = \varepsilon\text{left-}\varepsilon\text{right}$. We know that electromagnetic radiations usually consist of both electric and magnetic field oscillations that are perpendicular to each other. For instance, the oscillations of linearly polarized are usually confined to the single plane and all the polarized light are described as the sum of linear states at right angles to each other. Whereas, the circularized light occurs in case the direction of the electric field vector rotates in propagating direction. The absorption of the circularly polarized light by optically active molecules will be in one direction. The circularized polarized can exist in two directions, viz., left or right in the case, if the vector rotates anticlockwise down the axis of propagation the light is supposed to be left circularly polarized (LCP). In case the light rotates in a clockwise direction, the light is right circularly polarized (RCP). One must keep in consideration that if both RCP and LCP are of the same amplitude, they are super-imposed with one another yielding linear polarized light. The differentially absorbed left and right light is hence measured and quantified for analysis purposes. For instance, UV CD is employed for determining the secondary structures of proteins. Whereas, vibrational CD and IR CD are used to investigate the structure of small proteins, nucleic acids, and other organic molecules. Moreover, the CD based on UV/Vis is used to investigate the charge transfer transition in the case of metal—protein complexes.

6.1 Basic workflow of CD

The interaction of circularly polarized light is maximum with the optically active chiral molecules that absorb LCP and RCP in different amounts as given in the below equation.

$$\Delta A = Al - Ar$$

The absorption of light results in a change in the amplitude of the wave. In the case of chiral samples, the molar absorptive of LCP and RCP are different and this differential results in different amplitudes resulting in an elliptically polarized wave. The technique follows single bean of UV absorption that needs an electrooptic modulator and detector for measurements. The left and right circularly polarized light is passed through the analyte in an alternating fashion. Later phenomenon is obtained by an electro-optic modulator, which is actually a crystal that transmits left or right-handed polarized components of linearly polarized light, which grossly depends on the electric field applied by the alternating current. The voltage is produced by the photomultiplier detector almost equal to the ellipticity of the resultant beam emerging from the sample under investigation. In addition, the light source continuously flushes with nitrogen to further avoid ozone formation.

6.2 Applications

- Helps in detecting the sensitivity to the chirality (handedness) of molecules.
- It aids in absolute stereochemistry
- Detection of enantiomeric composition
- Racemization of compounds
- Enantiomeric differentiation is also analyzed
- Molecular interactions and conformation of proteins and other biomolecules are also confirmed
- CD is also used to study the mechanism of action by drugs
- Moreover, the affinity of each enantiomer of chiral drugs with target proteins and other biological macromolecules is also analyzed
- It helps in checking the protein drug quality control
- In combination with High Performance Liquid Chromatography (HPLC), CD helps in the content determination of chiral drugs

7. Conclusion

Biotechniques are very critical for studying the characteristics of proteins and other biomolecules for diverse applications in living as well as nonliving systems. Techniques such as ultraviolet and visible light spectroscopy (UV), fluorescence spectroscopy, CD, NMR, and XRD are frequently used for the analysis of biomolecules (Hofmann & Clokie, 2018). These techniques help to study biological molecules, their structure, interactions, and other properties. This chapter is aimed at investigating the application of biomolecules by using the previously mentioned techniques.

References

Albani, J. (2007). *Principles and applications of fluorescence spectroscopy.* Blackwell Science.

Alexandra, et al. (2004). Virus particles monitored by fluorescence spectroscopy: A potential detection assay for macromolecular assembly. *Photochemistry and Photobiology, 80*(1), 41–46.

Alexandra, et al. (2005). Virus particles and receptor interaction monitored by fluorescence spectroscopy. *Photochemistry and Photobiology, 1*(4), 879–883.

Ardui, S., Ameur, A., Vermeesch, J. R., & Hestand, M. S. (2018). Single-molecule real-time (SMRT) sequencing comes of age: Applications and utilities for medical diagnostics. *Nucleic Acids Research, 46*(5), 2159–2168.

Ayad, M. F., & Magdy, N. (2015). Application of new spectrofluorometric techniques for determination of atorvastatin and ezetimibe in the combined tablet dosage form. *Chemical and Pharmaceutical Bulletin, 63*(6), 443–449.

Bunaciu, A. A., Udriştioiu, E., Aboul-enein, H. Y., Bunaciu, A. A., Udriştioiu, E., Aboul-enein, H. Y., Bunaciu, A. A., & S, E. G. U. (2015). X-ray diffraction: Instrumentation and applications X-ray diffraction: Instrumentation and applications. *Critical Reviews in Analytical Chemistry, 45*, 289–299. https://doi.org/10.1080/10408347.2014.949616

Dahm, K. G., Van Straaten, C. M., Munakata-Marr, J., & Drewes, J. E. (2012). Identifying well contamination through the use of 3-D fluorescence spectroscopy to classify coalbed methane-produced water. *Environmental Science & Technology, 47*(1), 649–656.

Gruner, S. M. (1994). X-ray detectors for macromolecular crystallography. *Current Opinion in Structural Biology, 4*, 765–769.

Hofmann, A., & Clokie, S. (Eds.). (2018). *Wilson and Walker's principles and techniques of biochemistry and molecular biology* (8th ed.). Cambridge University Press. https://doi.org/10.1017/9781316677056

Kohli, R. (2012). Methods for monitoring and measuring cleanliness of surfaces. In *Developments in surface contamination and cleaning: Detection, characterization, and analysis of contaminants* (pp. 107–178). Elsevier.

Lakowicz, J. R. (Ed.). (2013). *Principles of fluorescence spectroscopy.* Springer Science & Business Media.

Leenheer, J. A., & Croué, J.-P. (2003). Characterizing aquatic dissolved organic matter. *Environmental Science & Technology, 37*(1), 18A–26A. https://doi.org/10.1021/es032333c

Leslie, A. (1993). Data collection and processing. In L. Sawyer, N. Isaac, & S. Bailey (Eds.), *Proceedings of the CCP4 study weekend* (pp. 44–51). SERC Daresbury Laboratory.

Milori, D. M. B. P., Raynaud, M., Villas-Boas, P. R., Venâncio, A. L., Mounier, S., Bassanezi, R. B., & Redon, R. (2013). Identification of citrus varieties using laser-induced fluorescence spectroscopy (LIFS). *Computers and Electronics in Agriculture, 95*, 11−18.

Moy, J.-P. A. (1994). 200 mm input field, 5−80 keV detector based on an X-ray image intensifier and CCD camera. *Nuclear Instruments and Methods in Physics Research Section A: Accelerators, Spectrometers, Detectors and Associated Equipment Methods, A348*, 641−644.

Röcker, C., Pötzl, M., Zhang, F., Parak, W. J., & Nienhaus, G. U. (2009). A quantitative fluorescence study of protein monolayer formation on colloidal nanoparticles. *Nature Nanotechnology, 4*(9), 577.

Sharma, A., & Schulman, S. G. (1999). *Introduction to fluorescence spectroscopy* (Vol 13). Wiley-Interscience.

Smyth, M., Tate, J., Hoey, E., et al. (1995). Implications for viral uncoating from the structure of bovine enterovirus. *Nature Structural & Molecular Biology, 2*, 224−231.

Tansman, G. F., Kindstedt, P. S., & Hughes, J. M. (December 2014). Powder X-ray diffraction can differentiate between enantiomeric variants of calcium lactate pentahydrate crystal in cheese. *Journal of Dairy Science, 97*(12), 7354−7362. https://doi.org/10.3168/jds.2014-8277

Analysis of proteomes—III

---

1. Protein structure prediction

The term "proteome" was given by Marc R. Wilkins, in 1994 and is defined as the entire protein characterization in an organism translated from the gene/s (Aslam et al., 2017). Proteins play important role in regulating different molecular processes in an organism. The function of proteins is dependent upon their three-dimensional conformation. Protein structure predictions are one of the significant aims established by computational biology (Cooper et al., 2010; Dorn et al., 2014). A small change or alteration in the protein structure leads to altered functions and thus, causes adverse effects (Kumar & Shanker, 2018). During evolution, the structure of the protein remains mostly conserved rather than its sequences. Proteins that share similar sequences form an identical structure while as distantly related proteins fold into almost similar structures. The protein sequence is usually obtained from the NCBI or UniProt (Kuhlman & Bradley, 2019).

---

2. Protein secondary and tertiary structure predictions

In structural biology, protein secondary structure is defined by the general three-dimensional conformation of protein segments or the conformations of the polypeptide backbone of proteins (Robson, 2022). There are two important secondary structure conformations: α-helix (H) and β-strand (E), as suggested by Linus Pauling 60 years ago, and one irregular type: coil region (C). Secondary structure predictions are a set of approaches in the field of bioinformatics that assist to identify and characterize the secondary configurations of protein sequences based on their primary structures (Rashid et al., 2016). The earlier methods of structure predictions, in the year the 1960s and 1970s, focused on the identification of α-helical structures and depended mainly upon helix—coil transition models. To identify or characterize the three-dimensional protein structures, various methods have been introduced such as homology or template-based modeling or comparative, fold recognition, and ab—initio structure predictions (Dorn et al., 2014).

3. Homology or template-based modeling

Homology modeling refers to designing a three-dimensional (3D) structure of proteins with the use of a known homologous protein structure (Schmidt et al., 2014). It is been observed that protein structures remain more conserved than their sequences among homologs, while as sequences showing similarity below the 20% identity attain different structures (Jing et al., 2019). This type of modeling generates high-definition structure models if the target and templates are closely similar to each other. Homology modeling is a many-step approach including alignment of sequences, modification in structure, searching database, and evaluation of structure to develop a unique structure (Muhammed & Aki-Yalcin, 2019). The steps involved are as follows:

1. Template identification and initial alignment
2. Generation of model
3. Loop modeling
4. Model optimization and
5. Validation

3.1 Template identification and initial alignment

The important step is the determination of the most similar template structure when it is available. One of the easiest methods is based on serial sequence alignments assisted by database search approaches such as BLAST and FASTA (Sehgal et al., 2018). The more sensitive approach that depends upon multisequence alignment (PSI-BLAST) stepwise updates the position-specific matrix to consecutively identify distantly related homologs. This set of approaches has the potential to generate a number of significant templates for similar sequences and to determine templates for sequences that are distantly related (Haddad et al., 2020).

3.2 Generation of models

When an alignment and a template are given, the information of sequences present there is utilized to design three-dimension structure designs for the targets, which is shown as a series of Cartesian coordinates for every atom inside a protein (Sehgal et al., 2018; Zhang, 2008). The cartesian coordinate system is a system that designates every atom, particularly by a pair of numerical, which represents the specified distance from the two fixed perpendicularly spaced lines to a certain point. The various classes of model generation approaches have been given in the following:

3.2.1 Assembly of fragments

The actual approach of homology modeling is based upon the generation of the complete model from the preserved structural fragments that are observed in closely related structures (Wallner & Elofsson, 2005; Xiang, 2007). For instance, observation of serine proteases in mammals showed intense variations among the "core" areas in the structure that is conserved in all the conformations of various classes, and in variable regions that are mainly localized in the loops, containing the majority of the differences in the sequences. The current application of this type of method differs mainly with regions that do not remain reserved or that lack the template. These variable regions are often generated by the use of fragment libraries.

3.2.2 Matching of segments

The segment-matching approach separates the target into various small segments, and each segment is confirmed by its template from the Protein Databases (Jiang et al., 2001). Thus, the matching of sequences is usually done on the segments and not on the whole protein. The template selection for every segment depends upon the similarity in sequences, α carbon coordinates comparisons, and steric hindrances that arise due to the van der Waals radius of diverging atoms between the template and target.

3.3 Modeling of loops

The target sequences that are not in parallel with a particular template are designed by the loop modeling; these are more prone to major errors during designing and occur in higher frequencies if the template and target have the least sequence similarities (Xiang, 2007). The coordinates of unmatched regions show accuracy when copied simply from the known structures especially when the loop is longer than ten residues while as are more accurate when characterized by the loop modeling process. The side-chain dihedral angles may be usually estimated within 30 degrees for backbone structures, and the later angles that are found in the larger side chains mostly arginine and lysine are usually difficult to predict. Even small errors in the one dihedral angle may lead to more errors in the various positions of the side chain terminal atoms and often to those which have some functional importance, especially those that are positioned at the active sites.

3.4 Model assessment

The assessment of homology modeling is usually performed with the help of two approaches: energy calculations or statistical potentials (Hameduh et al., 2020). Both approaches help generate an energy estimate or an energy-like analog for the model being studied. The statistical potential approach is an experimental technique that depends upon the observed residue—residue contact recurrences among the proteins of various known structures in the Protein Database (Bepler & Berger, 2019). It assigns an energy score or a probability to the possible pair-wise interactions among amino acids and combines their interaction scores into one score for the whole model. These approaches also produce a residue—residue analysis and are sensitive to lower scoring areas in the designed model, although the model shows effective scores all over. These approaches emphasize the hydrophobic cores and hydrophilic amino acids exposed via solvents that are majorly present in globular proteins. Statistical potential examples include DOPE and Prosa. The statistical potentials are computationally very effective than the energy calculations (Rykunov & Fiser, 2010). Energy calculations aim to identify the interatomic interactions that account for the protein stabilities in the solution such as electrostatic and van der Waals interactions.

4. Protein threading

Threading of proteins is an approach to structure prediction, in which the prediction is done by aligning or placing the amino acids present in the target to a point in the known template and evaluating the target's fitness with the template (Peng & Xu, 2011). It works by utilizing the statistical sequence data relationship present in the protein database between the template and the unknown target. It is also called fold recognition as it is used to design those proteins that possess the same folds of known structures and differ from the homology modeling as it is used on the proteins that do not possess homology. The structure is built only when the target fits with the known template is done. It is based upon two main observations—first, the number of different folds is usually small such as about 1300, and second that the new designs submitted to the protein database from the last 3 years possess the same structural folds as the ones already present in the protein databank.

4.1 Software for protein threading

RAPTOR: a software of protein threading based upon integer programming. It employs probabilistic graphic models and statistical data for both multi- and single-template-based threading. It is nowadays replaced with RaptorX Software. It is particularly good at aligning the proteins with low sequence profiles and is free to the public.

HHpred: a threading server that is used for detection of homology, which is upon pair-wise comparisons of hidden Markov model.

MUSTER: a standard threading algorithm software that is based upon the sequence profile alignments and also combines the multistructure data that help in such sequence alignments.

Phyre: is a most known threading software that combines HHsearch with multitemplate modeling and ab initio modeling.

SPARKS X: it is a probabilistic software based upon a sequence-to-structure match between the predicted 1D-structural queries and their analogous native properties of different templates.

BioShell: a threading algorithm that uses the advanced profile-to-profile programming algorithm associated with the predicted secondary structures.

5. Ab initio protein structure predictions

The ab initio modeling is also known as de novo modeling, free modeling, or physics-based modeling (Lee et al., 2017). This method depends upon the thermodynamic hypothesis that was proposed by Anfinsen and states that the original structure correlates to the lowest free energy under some conditions. This approach is one of the most general methods used for the target proteins which possess folds along with random conformations. The various another ab initio approach available are TOUCHSTONE-II, ROSETTA, and the most widely used I-Tasser that is based on the Monte-Carlo algorithm (Lee et al., 2017). The ab initio approach involves the basic protocol that starts with the searching of various conformations based upon the basic amino acid sequences and leads to the generation of native folds. After the recognition, as well as prediction, is done, the assessment of the model is done to analyze the standards of the predicted structures. The ROSETTA involves small fragment identifications such as 3 and 9 mers among the databases that are constant with the preferred general sequences. These fragments are then assembled into a model with general features and followed by the model assessment by use of a score function from the decoy populations.

Models that attempted to explain protein foldings involve the following:

- **Assumption**: Proteins in their native configuration is found to assume the lowest free energy.
- **Models**: A hierarchical phenomenon in which the natural secondary structure attains a form followed by supersecondary structure later and continues until the whole polypeptide folds. The spontaneous merge of the polypeptide into the compact globular state is known as a molten globule.

Software list:

1. Homology and threading modeling software

Name	Method	Link	References
IntFOLD	An interface used for tertiary structure predictions, 3D-model quality assessments, predictions of domain, predictions of protein–ligand binding residues	http://www.reading.ac.uk/bioinf/IntFOLD/	McGuffin et al. (2019)
RaptorX	Remote homology detections, 3D protein modeling, binding site predictions	http://raptorx.uchicago.edu/	Källberg et al. (2014)
ESyPred3D	Alignment, detection of template, 3D modeling	http://www.unamur.be/sciences/biologie/urbm/bioinfo/esypred/	Lambert et al. (2002)
Biskit	Wrapping of external programs into an automatic workflow	http://biskit.pasteur.fr/	Majumder (2020)
Robetta	Rosetta homology modeling and ab initio fragment assembly	http://robetta.bakerlab.org/	Kim et al. (2004)
HHpred	Template detection, alignment, 3D modeling	http://www.protevo.eb.tuebingen.mpg.de/hhpred	Söding et al. (2005)
MODELLER	Standalone program especially in PythonandFortran	http://www.salilab.org/modeller/	Webb and Sali (2014)

Continued

Name	Method	Link	References
SWISS MODEL	General similarities or fragment assemblies	http:// swissmodel. expasy.org/	Waterhouse et al. (2018)
Phyre_/ _Phyre2	Distantly related template alignment and detection, 3D formulation of multitemplates, ab initio	http://www.sbg. bio.ic.ac.uk/ phyre2	Kelley and Sternberg (2009)

2. Ab Initio Structure predictions

Name	Description	Link	References
TrRosetta	It is an algorithm for the exact and sharp de-novo protein structure predictions. It models the protein structure based on direct energy minimizations. The restraints include interresidue distance and orientation distributions	https:// zhanglab. ccmb.med. umich.edu/ casp14	Du et al. (2021)
I-TASSER	Threading fragments and structure assemblies	http:// zhanglab. ccmb.med. umich.edu/ ITASSER/	Yang and Zhang, (2015)
Rosetta	Homology modeling and ab initio fragment assemblies	http://boinc. bakerlab.org/ rosetta/	Rohl et al. (2004)
Swiss model	General similarities or fragment assemblies	http:// swissmodel. expasy.org/	Waterhouse et al. (2018)

6. Conclusions

Protein structure prediction approaches are found to play important role in designing a complete structural proteome profile. Large experimental data will assist in future machine learning and help to lead an effective accuracy as well as speed due to the enhancement in the number of structural templates. Further, establishing protein interaction networks is important in determining the different cellular processes. Tertiary structure prediction is used to discover the whole network interactions and validate the molecular details important for interactions when proposed by various other approaches. Thus, we expect that the prediction of protein structure will help to identify the structure–function relation of proteins and assist in studying and establishing the protein interactions that will significantly help in the postgenomic era.

References

Aslam, B., Basit, M., Nisar, M. A., Khurshid, M., Rasool, M. H., et al. (2017). Proteomics: Technologies and their applications. *Journal of Chromatographic Science, 55*(2), 182–196.

Bepler, T., & Berger, B. (2019). *Learning protein sequence embeddings using information from structure.* arXiv preprint arXiv:1902.08661.

Cooper, S., Khatib, F., Treuille, A., Barbero, J., Lee, J., Beenen, M., Popović, Z., et al. (2010). Predicting protein structures with a multiplayer online game. *Nature, 466*(7307), 756–760.

Dorn, M., e Silva, M. B., Buriol, L. S., Lamb, L. C., et al. (2014). Three-dimensional protein structure prediction: Methods and computational strategies. *Computational Biology and Chemistry, 53*, 251–276.

Du, Z., Su, H., Wang, W., Ye, L., Wei, H., Peng, Z., Yang, J., et al. (2021). The trRosetta server for fast and accurate protein structure prediction. *Nature Protocols, 16*(12), 5634–5651.

Haddad, Y., Adam, V., & Heger, Z. (2020). Ten quick tips for homology modeling of high-resolution protein 3D structures. *PLoS Computational Biology, 16*(4), e1007449.

Hameduh, T., Haddad, Y., Adam, V., & Heger, Z. (2020). Homology modeling in the time of collective and artificial intelligence. *Computational and Structural Biotechnology Journal, 18*, 3494–3506.

Jiang, W., Baker, M. L., Ludtke, S. J., & Chiu, W. (2001). Bridging the information gap: Computational tools for intermediate resolution structure interpretation. *Journal of Molecular Biology, 308*(5), 1033–1044.

Jing, X., Dong, Q., Hong, D., & Lu, R. (2019). Amino acid encoding methods for protein sequences: A comprehensive review and assessment. *IEEE/ACM Transactions on Computational Biology and Bioinformatics, 17*(6), 1918–1931.

Källberg, M., Margaryan, G., Wang, S., Ma, J., & Xu, J. (2014). RaptorX server: A resource for templatebased protein structure modeling. *Protein Structure Prediction, 1137*, 17–27.

Kelley, L. A., & Sternberg, M. J. (2009). Protein structure prediction on the web: A case study using the Phyre server. *Nature Protocols, 4*(3), 363–371.

Kim, D. E., Chivian, D., & Baker, D. (2004). Protein structure prediction and analysis using the Robetta server. *Nucleic Acids Research, 32*(Suppl. 1_2), W526–W531.

Kuhlman, B., & Bradley, P. (2019). Advances in protein structure prediction and design. *Nature Reviews Molecular Cell Biology, 20*(11), 681–697.

Kumar, S., & Shanker, A. (2018). Bioinformatics Resources for the stress biology of plants. In *Biotic and abiotic stress tolerance in plants* (pp. 367–386). Springer.

Lambert, C., Leonard, N., De Bolle, X., & Depiereux, E. (2002). ESyPred3D: Prediction of proteins 3D structures. *Bioinformatics, 18*(9), 1250–1256.

Lee, J., Freddolino, P. L., & Zhang, Y. (2017). Ab initio protein structure prediction. In *From protein structure to function with bioinformatics* (pp. 3–35). Springer.

Majumder, P. (2020). Computational methods used in prediction of protein structure. In *Statistical modelling and machine learning principles for bioinformatics techniques, tools, and applications* (pp. 119–133). Springer.

McGuffin, L. J., Adiyaman, R., Maghrabi, A. H., Shuid, A. N., Brackenridge, D. A., Nealon, J. O., Philomina, L. S., et al. (2019). IntFOLD: An integrated web resource for high performance protein structure and function prediction. *Nucleic Acids Research, 47*(W1), W408–W413.

Muhammed, M. T., & Aki-Yalcin, E. (2019). Homology modeling in drug discovery: Overview, current applications, and future perspectives. *Chemical Biology & Drug Design, 93*(1), 12–20.

Peng, J., & Xu, J. (2011). A multiple-template approach to protein threading. *Proteins: Structure, Function, and Bioinformatics, 79*(6), 1930–1939.

Rashid, S., Saraswathi, S., Kloczkowski, A., Sundaram, S., & Kolinski, A. (2016). Protein secondary structure prediction using a small training set (compact model) combined with a complex-valued neural network approach. *BMC Bioinformatics, 17*(1), 1–18.

Robson, B. (2022). De novo protein folding on computers. Benefits and challenges. *Computers in Biology and Medicine*, 105292.

Rohl, C. A., Strauss, C. E., Misura, K. M., & Baker, D. (2004). Protein structure prediction using Rosetta. In , *383. Methods in enzymology* (pp. 66–93). Academic Press.

Rykunov, D., & Fiser, A. (2010). New statistical potential for quality assessment of protein models and a survey of energy functions. *BMC Bioinformatics, 11*(1), 1–11.

Schmidt, T., Bergner, A., & Schwede, T. (2014). Modelling three-dimensional protein structures for applications in drug design. *Drug Discovery Today, 19*(7), 890–897.

Sehgal, S. A., Mirza, A. H., Tahir, R. A., Mir, A., et al. (2018). *Quick guideline for computational drug design*. Bentham Science Publishers.

Söding, J., Biegert, A., & Lupas, A. N. (2005). The HHpred interactive server for protein homology detection and structure prediction. *Nucleic Acids Research, 33*(Suppl. 1_2), W244–W248.

Wallner, B., & Elofsson, A. (2005). All are not equal: A benchmark of different homology modeling programs. *Protein Science, 14*(5), 1315–1327.

Waterhouse, A., Bertoni, M., Bienert, S., Studer, G., Tauriello, G., Gumienny, R., Schwede, T., et al. (2018). SWISS-MODEL: Homology modelling of protein structures and complexes. *Nucleic Acids Research, 46*(W1), W296–W303.

Webb, B., & Sali, A. (2014). Protein structure modeling with MODELLER. *Methods Mol Biol., 1137*, 1–15. https://doi.org/10.1007/978-1-4939-0366-5_1. PMID: 24573470.

Xiang, Z. (2007). Homology-based modeling of protein structure. In *Computational methods for protein structure prediction and modeling* (pp. 319–357). Springer.

Yang, J., & Zhang, Y. (2015). I-TASSER server: new development for protein structure and function predictions. *Nucleic Acids Res, 43*(W1), W174–W181. https://doi.org/10.1093/nar/gkv342. Epub 2015 Apr 16. PMID: 25883148; PMCID: PMC4489253.

Zhang, Y. (2008). Progress and challenges in protein structure prediction. *Current Opinion in Structural Biology, 18*(3), 342–348.

Analysis of proteomes—IV

1. Introduction

Proteome analysis refers to the separation, identification, and measurement of the entire protein complement expressed by a genome, cell, or tissue. Protein—protein interactions (PPIs) govern a variety of biological activities, including cell-to-cell communication, and metabolic and developmental regulation. Protein—protein interaction is soon becoming one of system biology's most significant tasks. PPIs can be classified in a variety of ways depending on their structural and functional characteristics. Based on their interaction surface, they might be homo- or hetero-oligomeric; compulsory or nonobligatory; and transient or permanent in terms of persistence (Acuner Ozbabacan et al., 2011; Poluri et al., 2021). A specific PPI could be a combination of these three different pairs. Persistent connections would result in a stable protein complex, whereas transient contacts would result in signaling pathways. Understanding cell biology will require characterizing the interactions of proteins in a specific proteome. Phizicky and Fields have identified the significant properties of PPIs (Fig. 9.1).

2. Yeast 2-hybrid system

The yeast 2-hybrid system (Y2H) approach is a method for detecting PPIs in vivo. The Y2H assay needs the use of two protein domains, each with a specific function:

1. a DNA binding domain that aids in DNA binding and
2. an activation domain (AD) that aids in DNA transcription activation.

Principles of Genomics and Proteomics. https://doi.org/10.1016/B978-0-323-99045-5.00008-2

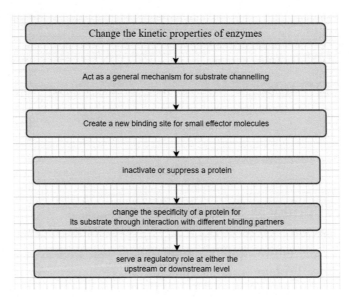

Change the kinetic properties of enzymes

↓

Act as a general mechanism for substrate channelling

↓

Create a new binding site for small effector molecules

↓

inactivate or suppress a protein

↓

change the specificity of a protein for
its substrate through interaction with different binding partners

↓

serve a regulatory role at either the
upstream or downstream level

FIGURE 9.1

Methods for detecting protein—protein interactions are classified into three types: In vitro, in vivo, and in silico. In vitro techniques are used to perform a process in a controlled environment outside of a living species. In vitro techniques for PPI identification include tandem affinity purification, affinity chromatography, coimmunoprecipitation, protein arrays, protein fragment complementation, yeast 2-hybrid, phage display, X-ray crystallography, and NMR spectroscopy.

Both domains are required for reporter gene transcription to occur. Y2H analysis can directly detect PPI between protein pairs (Fig. 9.2). In a Y2H experiment, interacting proteins must be targeted to the nucleus because proteins that are less likely to be present in the nucleus are rejected due to their incapacity to activate reporter genes. Proteins that need posttranslational modifications to function are unlikely to behave or interact correctly. The Y2H method, as well as other in vitro and in vivo approaches, resulted in the widespread development of useful tools for detecting PPIs between specific proteins that can occur in a variety of combinations (Lopez & Mukhtar, 2017; Paiano et al., 2019). This is a powerful tool for researchers, and it is typically used with one or two additional ways to explore

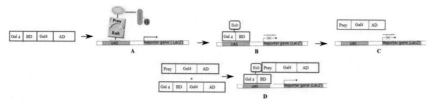

FIGURE 9.2

The Y2H assay uses two protein domains. (A) Regular transcription of receptor gene. (B) One fusion protein only (Gal4 + BD + Bait) = no transcription. (C) One fusion protein only (Gal4 + AD + Prey) = no transcription. (D) Two fusion proteins interacting with Bait and Prey.

the variety of interactions that occur in cells. The test is straightforward to perform and yields high-quality results in a short amount of time.

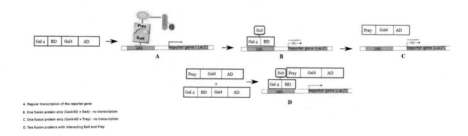

A. Regular transcription of the reporter gene

B. One fusion protein only (Gal4-BD + Bait) – no transcription

C. One fusion protein only (Gal4-AD + Prey) – no transcription

D. Two fusion proteins with interacting Bait and Prey

3. **Principle**

- The Y2H test is based on the binding of a particular transcription factor to a reporter gene (such as lacZ or Green fluorescent protein (GFP)).
- There are two domains in the transcription factor: a DNA-binding domain (BD) and an AD.
- The Prey is the protein library fused with the AD, while the Bait is the query protein of interest fused with the BD.
- To activate the reporter gene, a transcriptional unit must be present at the gene locus, which is only possible if Bait and Prey interact.

4. Advantages and limitations

Because of its simple methodology and quick turnaround time, Y2H is an effective technology for identifying protein interactions. As a result, the technique's throughput may be greatly increased to scan the full proteome (Smith, 1985). Furthermore, the approach has been applied to explore organism-specific interactions in additional model organisms (Ledsgaard et al., 2018).

5. Limitations of Y2H assays

1. For the reporter gene to be activated, interaction must occur in the cell's nucleus. Even if proteins interact directly, proteins that are sequestered in different cellular compartments may not cause a beneficial interaction. This can be avoided by utilizing a split ubiquitin 2-hybrid system or doing the experiment in bacteria that lack a nucleus.
2. The test has a high rate of false positive and false negative interactions. This is an essential reason why any interactions should be validated using additional approaches such as coimmunoprecipitation.
3. Overexpression of recombinant fusion protein, which occurs in the majority of Y2H studies, may result in erroneous interaction results. Furthermore, the fusing of AD/DB domains to query proteins may have an effect on query protein function in vivo.

6. Phage display

George P. Smith described phage display in 1985 as a means for identifying a gene against which he had produced antibodies (Smith, 1985). Greg Winter and John McCafferty of the Laboratory of Molecular Biology in Cambridge, UK, and Richard Lerner and Carlos F. Barbas of The Scripps Research Institute in the United States each employed phage display to construct vast libraries of completely human antibody sequences. This research paved the way for the creation of human antibody-based medicines. There are several types of phage display libraries available today, including peptide libraries, protein libraries, and antibody libraries. Phage display is a laboratory platform that enables scientists to analyze large-scale protein interactions and identify proteins with the best affinity for certain targets. Phage display technology is a technique for discovering ligands for proteins and other macromolecules in vitro. The capacity to produce peptide or protein sequences as fusions to bacteriophage coat proteins is crucial to phage display technology (Hess & Jewell, 2020; Ledsgaard et al., 2018). Libraries of phage-displayed peptides or proteins are therefore physically attached to their encoding nucleic acid, enabling for the selection of binding partners for a wide range of target types by repetitive rounds of in vitro panning and amplification followed by DNA sequencing. In a matter of days, libraries with over a billion members may be screened, providing an efficient alternative to more traditional methods of epitope mapping, receptor-ligand discovery, or protein evolution.

The primary benefit of phage display is that it allows for the identification of target-binding proteins from a library of millions of distinct proteins without the need to screen each molecule individually. This allows for the screening of billions of proteins per week. Phage display, by tying a specific protein to its encoding gene, also allows for the easy identification of binding protein-coding sequences. These can be saved, magnified, or otherwise handled. Phage display is an important tool for fundamental scientific study as well as the creation of novel medications and vaccines. The method has been very useful in the development of safer and more effective monoclonal antibody medicines. Phage display libraries constructed completely of human antibody sequences, for example, have enabled the production of totally human antibodies (Ledsgaard et al., 2022; Nagano & Tsutsumi, 2021).

7. Application

The phage display approach is based on the creation of a library of millions of bacteriophages that have been genetically modified to show various peptides or proteins on their surface. This is accomplished by introducing a gene encoding the desired protein into the phage's protein shell, so establishing a direct physical connection between DNA sequences and their encoding proteins. The modification's goal is to create a molecule that can replicate a natural modulator within the biological process. For the purpose of phage display, many kinds of phages are utilized. The most common are filamentous bacteriophages. The genetically engineered phages are compiled into a library to be used as a screening platform for proteins, peptides, and DNA sequences. The phage-display library is screened by adding it to the wells of a microtiter plate containing immobilized target proteins or DNA sequences. The plate is then incubated for a period of time to allow the phages to bond with the target of interest before being rinsed to remove any nonbinding phages. Any remaining phages connected to the wells are then removed and injected into other bacteria for reproduction. The cycle is continued until only phage-displaying proteins with great target specificity remain (Anand et al., 2021; Rahbarnia et al., 2017; Zambrano-Mila et al., 2020). Both hybridoma and phage display is a critical technology for producing monoclonal antibodies. However, each approach has limits; therefore, researchers should assess their benefits and drawbacks using Fig. 9.3A and B.

8. Protein chips

DNA microarrays form the basis of protein arrays or protein microarrays or protein chips, and this technology has been successful in allowing high-throughput proteomic investigations and other large-scale genomic studies. This technology has allowed researchers to delve into and investigate the entire proteome of a cell to automate and parallelize protein investigations rapidly and efficiently. Various steps involved in this technology include examining the protein chip levels followed by an analysis of the extensive data quantity obtained through computer software. Despite the large success of the technology, there are some obstacles like a large number of proteins present in cells, fluctuating or inconsistent protein concentration, and apprehending protein-specific probes (Berrade et al., 2011; Zambrano-Mila et al., 2020).

The application and use of protein chip technology in discovering diagnostic disease biomarkers, antibody detection in a sample, and

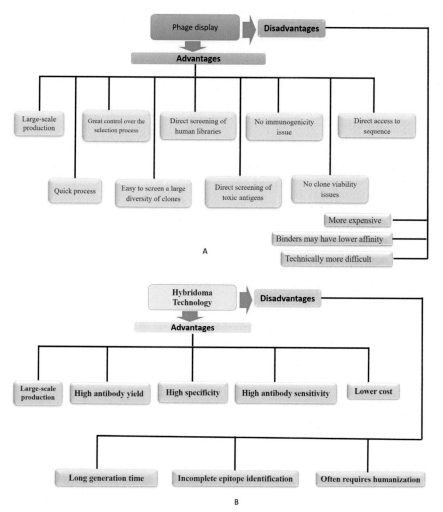

FIGURE 9.3

Advantages and disadvantages of phage display and hybridoma technology.

studying protein—protein interactions have opened gateways for their utilization in research related to cancer studies, diagnostic medical testing, and proteomics. Due to their widespread application, protein microarrays have become essential tools in the fields of biochemistry and molecular biology. Analytical and functional protein arrays are the two types of protein microarrays available. The former is the most robust multiform detection method, particularly antibody arrays while the latter is being used in various aspects of biological research, such as immunological responses, and investigations on interactions between proteins and metabolic activities. However, both types have contributed significantly in terms of applicability, specificity, and sensitivity. The specific uses of analytical protein arrays include detection of expression levels of proteins, identification of biomarkers, patient diagnosis, and assessment of environmental/food safety (Aslam et al., 2017; Duarte & Blackburn, 2017). Various purposes of using functional protein microarrays include the following:

I. Activity-based protein profile probing such as protein interactions between lipids, DNA, peptide, and drug.
II. Identification of substrates for enzymes; and
III. Profiling immunological responses. Fig. 9.4 depicts the applicability of both types of protein arrays.

9. Applications of protein microarrays
9.1 Analytical microarrays

Among the analytical arrays, the most peculiar type is the antibody array which involves the distribution of antibodies at great density

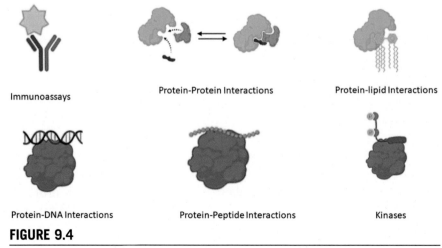

Immunoassays

Protein-Protein Interactions

Protein-lipid Interactions

Protein-DNA Interactions

Protein-Peptide Interactions

Kinases

FIGURE 9.4

Applicability of protein array.

on surfaces of glass but the development of antibodies capable of recognizing highly specific proteins with high-throughput affinity is the major problem of antibody microarrays. Researchers have been compelled to look for an alternative to this technology because of the labor-intensive and time-consuming process of developing monoclonal antibodies. However, to speed up the process of manufacturing high-specificity antibodies, various technologies like systematic evolution of ligands by exponential enrichment, phage antibody display, messenger RNA display, ribosome display, and affibody display have been developed. The above-mentioned procedures involve creating an extensive repository of applicable areas with potential binding activity chosen through innumerable series of affinity purification. The affinity of the consequent clone binding candidates can further be increased through maturation techniques. However, efforts are being made to develop a perfect selection system that should not only be robust, rapid, sensitive, and inexpensive but also automated and reduced (Cahill, 2001; Haab, 2001; Stoll et al., 2002; Templin et al., 2002).

9.2 Functional protein microarrays

This type of microarray has favored many aspects of detection-based biological research, including protein interactions with lipids, DNA, drug, and peptide. Despite our efforts to present all the key uses of functional protein arrays, it is difficult to encompass all the cases where they have been employed. As a result, we choose to focus the majority of our discussion on microarrays of yeast proteome.

9.2.1 Protein–protein and protein–lipid interactions

The application of proteome microarrays for the first time comprising around 5800 isolated yeast proteins making up to 85% of the yeast proteome was described by Zhu and colleagues (Zhu et al., 2001). These chips when prepared with biotinylated calmodulin were utilized for examining various protein–protein interactions and detecting some novel binding partners. Investigations with respect to interactions between proteins and several phospholipids known to function as secondary messengers were carried out using protein arrays. Probes of biotinylated liposomes containing specific phospholipids detected over 150 proteins binding to phospholipids with a diverse set binding to lipid vesicles out of which 50% have been accounted to have membrane association.

9.2.2 Protein–DNA interactions

The aforementioned group of researchers conducted screening for novel binding capabilities of DNA via protein chips using fluorescence-tagged genomic DNA of yeast in a later paper (Ptacek et al., 2005). There were 200 proteins detected that bind DNA in a reproducible manner. Half of these have never been proven to have binding activity with DNA before and have been found to belong to a broad range of functional groups. The Discovery of a novel DNA binding protein, that is, Arg5,6 was the most startling finding in this category. This gene has been found to encode two enzymes in mitochondria which act as mediators of two critical stages in ornithine production (a precursor to arginine). Supplementary research studies demonstrated a connection between this enzyme and certain loci (mitochondrial) in vivo which was employed in the establishment of a motif (DNA binding) for the said enzyme due to which a novel activity related to DNA binding to a protein was identified which was well-characterized, releasing the latest role of that protein.

9.2.3 Protein–drug interactions

Using protein arrays to identify targets and develop new drugs has also shown great potential. Since the binding pattern of an intriguing drug may be examined simultaneously over the full proteome, it is possible to evaluate the specificity or side effects of the medication. Additionally, this information ought to provide crucial pointers for improving drug design (Hall et al., 2004). By investigating proteome chips of yeast with biotinylated small molecule inhibitors of rapamycin (SMIRs) to discover genetic modifiers of the target of the rapamycin signaling network, Huang and coworkers (Huang et al., 2004a, 2004b; Lueking et al., 2003) used protein microarrays to find pharmacological targets. They identified proteins that the SMIRs may target therapeutically and established the identity of a previously unidentified protein as a legitimate SMIR target.

10. Conclusion and future prospective

The technological simplicity and methodological diversity of Y2H have demonstrated its capacity to quickly generate a considerable number of reliable protein—protein interaction data. Current Y2H technologies, especially those relying on split proteins, enable access to approximately the whole cellular proteome and the study of protein—protein interactions in their natural physiological regions. In that it reveals direct contacts in addition to lower affinity and transitory interactions, Y2H is extremely supportive to modern Affinity purification-mass spectrometry (AP/MS) techniques. It will take a lot of work to model the complex and dynamic network of interactions within a cell. Large interaction networks can only be accurately described using a combination of approaches, such as Y2H, MS, and bioinformatics.

One of the most essential tools of the applied part of biological research is the technology of phage display because of its swift, effective, and affordable application for examining interactions between protein—protein, sites for receptor binding, and discovering epitopes, mimotopes, functional, and accessible antigen areas. This technology has drawn the attention of numerous researchers due to its enormous versatility, enabling the use of phage display-based techniques to examine disease states, improve the procedures for diagnosis, and develop powerful remedial medications and vaccines. The integration of this technology in the formulation of vaccines and their distribution will be covered in the second section.

The efficiency of protein microarray technology for multimodal detection and proteomics research has been established. Functional protein microarrays have seen a significant rise in activities, and analytical protein microarrays have developed femtomolar responsiveness. Among the most effective diagnostic and high-throughput biological tools now in use seems to be protein microarrays. Quantitative analysis will rely primarily on our capability to produce huge quantities of antibodies, proteins, or their substitutes of high quality stimulating the use of this technique to more model species.

References

Acuner Ozbabacan, S. E., Engin, H. B., Gursoy, A., & Keskin, O. (2011). Transient protein-protein interactions. *Protein Engineering, Design, and Selection, 24*(9), 635–648.

Anand, T., Virmani, N., Bera, B. C., Vaid, R. K., Vashisth, M., Bardajatya, P., … Tripathi, B. N. (2021). Phage display technique as a tool for diagnosis and antibody selection for coronaviruses. *Current Microbiology, 78*(4), 1124–1134.

Aslam, B., Basit, M., Nisar, M. A., Khurshid, M., & Rasool, M. H. (2017). Proteomics: Technologies and their applications. *Journal of Chromatographic Science, 55*(2), 182–196.

Berrade, L., Garcia, A. E., & Camarero, J. A. (2011). Protein microarrays: Novel developments and applications. *Pharmaceutical Research, 28*, 1480–1499. https://doi.org/10.1007/s11095-010-0325-1

Cahill, D. J. (2001). Protein and antibody arrays and their medical applications. *Journal of Immunological Methods, 250*, 81–91 (Crossref. PubMed).

Duarte, J. G., & Blackburn, J. M. (2017). Advances in the development of human protein microarrays. *Expert Review of Proteomics, 14*(7), 627–641.

Haab, B. B. (2001). Advances in protein microarray technology for protein expression and interaction profiling. *Current Opinion in Drug Discovery and Development, 4*, 116–123 (PubMed).

Hall, D. A., Zhu, H., Zhu, X., Royce, T., Gerstein, M., & Snyder, M. (2004). Regulation of gene expression by a metabolic enzyme. *Science, 306*, 482–484 (Crossref. PubMed).

Hess, K. L., & Jewell, C. M. (2020). Phage display as a tool for vaccine and immunotherapy development. *Bioengineering & Translational Medicine, 5*(1), e10142.

Huang, Y. H., Li, D., Winoto, A., & Robey, E. A. (2004). Distinct transcriptional programs in thymocytes responding to T cell recep tor, Notch, and positive selection signals. *Proceedings of the National Academy of Sciences of the United States of America, 101*, 4936–4941 (Crossref. PubMed).

Huang, J., Zhu, H., Haggarty, S. J., Spring, D. R., Hwang, H., Jin, F., Snyder, M., & Schreiber, S. L. (2004). Finding new components of the target of rapamycin (TOR) signaling network through chemical ge netics and proteome chips. *Proceedings of the National Academy of Sciences of the United States of America, 101*, 16594–16599 (Crossref. PubMed).

Ledsgaard, L., Kilstrup, M., Karatt-Vellatt, A., McCafferty, J., & Laustsen, A. H. (2018). Basics of antibody phage display technology. *Toxins, 10*(6), 236.

Ledsgaard, L., Ljungars, A., Rimbault, C., Sørensen, C. V., Tulika, T., Wade, J., … Laustsen, A. H. (2022). Advances in antibody phage display technology. *Drug Discovery Today, 27*(8), 2151−2169. https://doi.org/10.1016/j.drudis.2022.05.002

Lopez, J., & Mukhtar, M. S. (2017). Mapping protein-protein interaction using high-throughput yeast 2-hybrid. In *Plant genomics* (pp. 217−230). New York, NY: Humana Press.

Lueking, A., Possling, A., Huber, O., Beveridge, A., Horn, M., Eickhoff, H., Schuchardt, J., Lehrach, H., et al. (2003). A nonredundant human protein chip for antibody screening and serum profiling. *Molecular and Cellular Proteomics, 2*, 1342−1349 (Crossref. PubMed).

Nagano, K., & Tsutsumi, Y. (2021). Phage display technology as a powerful platform for antibody drug discovery. *Viruses, 13*(2), 178.

Paiano, A., Margiotta, A., De Luca, M., & Bucci, C. (2019). Yeast two-hybrid assay to identify interacting proteins. *Current Protocols in Protein Science, 95*(1), e70.

Poluri, K. M., Gulati, K., & Sarkar, S. (2021). Prediction, analysis, visualization, and storage of protein-protein interactions using computational approaches. In *Protein-protein interactions* (pp. 265−346). Singapore: Springer.

Ptacek, J., Devgan, G., Michaud, G., Zhu, H., Zhu, X., Fasolo, J., Guo, H., Jona, G., et al. (2005). Global analysis of protein phosphorylation in yeast. *Nature, 438*, 679−684 (Crossref. PubMed).

Rahbarnia, L., Farajnia, S., Babaei, H., Majidi, J., Veisi, K., Ahmadzadeh, V., & Akbari, B. (2017). Evolution of phage display technology: From discovery to application. *Journal of Drug Targeting, 25*(3), 216−224.

Smith, G. P. (1985). Filamentous fusion phage: Novel expression vectors that display cloned antigens on the virion surface. *Science, 228*, 1315−1317. https://doi.org/10.1126/science.4001944

Stoll, D., Templin, M. F., Schrenk, M., Traub, P. C., Vohringer, C. F., & Joos, T. O. (2002). Protein microarray technology. *Frontiers in Bioscience, 7*, c13−c32 (Crossref. PubMed).

Templin, M. F., Stoll, D., Schrenk, M., Traub, P. C., Vohringer, C. F., & Joos, T. O. (2002). Protein microarray technology. *Trends in Biotechnology, 20*, 160−166 (Crossref. PubMed).

Zambrano-Mila, M. S., Blacio, K. E. S., & Vispo, N. S. (2020). Peptide phage display: Molecular principles and biomedical applications. *Therapeutic Innovation & Regulatory Science, 54*(2), 308−317.

Zhu, H., Bilgin, M., Bangham, R., Hall, D., Casamayor, A., Bertone, P., Lan, N., Jansen, R., Bidlingmaier, S., Houfek, T., Mitchell, T., Miller, P., Dean, R. A., Gerstein, M., & Snyder, M. (2001). Global analysis of protein activities using proteome chips. *Science, 293*, 2101−2105. https://doi.org/10.1126/science.1062191 (Crossref. PubMed).

Beyond genomics and proteomics

10

1. Introduction

Omics technologies are a collection of high throughput analyses based on studies such as genomics, transcriptomics, metabolomics, lipidomics, and proteomics to provide an in-depth profile of a total number of coding and noncoding sequences, RNAs, and proteins from a cell or organism. The studies of genomics unravel the realistic details regarding the structure of coding and noncoding regions in a genome. The last few decades of the genomic era led to the deciphering of huge information pertaining to the construction and composition of genomes by the application of genomic technologies. By employing automated sequencing technologies and computer algorithms, the genomes were used to sequence and assemble large amount of nucleotide sequences of humans and several model organisms. On the other hand, transcriptomics aids in studying the total RNA transcripts coding (mRNA) and noncoding RNAs categorized as short noncoding RNAs and long noncoding RNAs found in a cell or organism. Several techniques have been applied to study the transcriptomes of the system such as RNAseq. and next-generation sequencing (NGS) as detailed in the proceeding sections. Proteomics deals with the investigation of the total protein content of a cell or organism and their structure and fontina dynamics to investigate the expression, regulation, and metabolic role of proteins. Other omics fields including metabolomics, ionomics, interactomics, and metagenomics are employed to study the total metabolite content, ions, and protein—proteins interaction and study microbiomes from environmental samples or other living systems, respectively. This chapter is aimed give a detailed background of all these omics technologies, workflow, their principle,

Principles of Genomics and Proteomics. https://doi.org/10.1016/B978-0-323-99045-5.00002-1

and applications in the field of life sciences. At the end of this chapter, we have also provided an overview of connectomics, to understand the expanding dimensions of omics technologies in the field of neurobiology. The proceeding section of this chapter will discuss all the mentioned omics approaches and their respective applications.

2. Transcriptomics

The study of transcriptomes, that is, a complete set of RNA molecules transcribed from a cell, tissues, organs, or organism from the genomes, assessed by high throughput technologies such as microarray analysis under specific conditions or circumstances. The comparative analysis of transcriptomes from one cell to another cell or tissues or organisms under specific conditions helps to elucidate the adaptations, status of health, and effect of exogenous and endogenous factors. The key technologies employed in transcriptomics include microarray analysis used to quantity the total RNA transcripts and RNA sequencing (RNA-Seq) employing high throughput sequencing to sequence all the transcripts. The first human transcriptome was published in 1991, which reported 609 mRNA sequences obtained from the human brain (Kelley et al., 1991). In addition, two human transcriptomes reporting millions of RNA transcripts covering 16,000 genes were published in 2008 (Pan et al., 2008; Sultan et al., 2008) and transcriptomes of several hundreds of humans were published in 2015 (Melé et al., 2015). Transcriptomics helps to measure the regulation of genes by studying the expression analysis of genes in cells and tissues in different conditions and time points.

2.1 Technological intervention

Several critical steps are operated for transcriptome analysis in combination with the two important high throughput technologies including microarrays and RNA-Seq developed during mid of the 1990s (Wang et al., 2009) and 2000s (Nelson, 2001) (Fig. 10.1).

2.1.1 Microarrays

In 1995, the first microarrays, involving the hybridization of the transcripts with the array of immobilized probes, were a major breakthrough in transcriptomic technology (Pozhitkov et al., 2007;

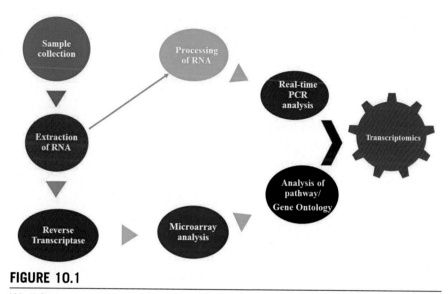

FIGURE 10.1

Workflow operated for accomplishing the transcriptome of a cell/tissue or organism.

Schena et al., 1995). Thousands of transcripts can be analyzed by microarray technology to reduce cost and time (Heller, 2002). The Affymetrix (Santa Clara, California) high-density arrays and spotted oligonucleotide arrays were the mainstream transcriptional profiling methods employed in the late 2000s (Dan Corlan, 2004; Nelson, 2001). Since the 2000s onwards, advances in microarray technology with respect to improvement in probe specificity led to test a large number of transcripts in a single array. In particular, fluoresce detection enhanced the accuracy in measurement and sensitivity for transcripts with low abundance (Ambroise et al., 2005; Pozhitkov et al., 2007).

The basic principle of microarrays lies in the arrangement of short nucleotide oligomers also known as probes arrayed on a solid support usually glass (Romanov et al., 2014). The abundance of transcripts is determined by the hybridization of these probes with fluorescently labeled transcripts (Barbulovic-Nad et al., 2006). The intensity of fluorescence at each spot in the microarray chip indicates the abundance of the transcripts (Barbulovic-Nad et al., 2006). It is critical to have prior knowledge of annotated genome sequences or expressed sequence tags to design a probe for microarrays.

Microarrays are manufactured by using nanofabrication and micro-fabrication techniques. Based on probe density, the microarrays are classified into the following:

- Low-density spotted arrays (Schena et al., 1995).
- High-density short probe arrays (Schena et al., 1995).

In the case of spotted low-density arrays, picolitre drops of purified cDNAs arrayed are arrayed on the glass chips (Auburn et al., 2005). Whereas, high-density arrays use a single channel and each of the samples is hybridized and detected individually to calculate the transcript abundance (Lockhart et al., 1996). The popular microarray chip Affymetrix GeneChip array uses high-density arrays embedded with 25 nucleotide probes to assay each transcript to analyze one gene (Irizarry et al., 2003). NimbleGen arrays (Pleasanton, CA), manufactured by maskless photochemistry method, are another set of high-density microarrays, consisting of 45–85 nucleotide long probes to accomplish through throughput expression analysis (Selzer et al., 2005).

2.1.2 RNA-Seq

The transcriptome analyzing technique, which employs NGS to unravel the presence and quantification of total transcripts in the given biological samples in given time and conditions (Chu & Corey, 2012). This technique involves the sequencing of cDNA transcripts and the abundance is derived from the total number of counts from each of the transcripts. Therefore, it is evident that RNA-Seq is largely dependent on high throughput sequencing technologies (Morozova et al., 2009; Wang et al., 2009). The technique facilitates the analysis of transcripts with spliced products, posttranscriptionally modified transcripts, mutations, and gene fusions expression analysis as per time and conditions (Maher et al., 2009).

The RNA-Seq transcriptomic technique is a high-throughput sequencing technology used to analyze RNA transcripts in combination with computational methods (Ozsolak & Milos, 2011). Depending on the method of sequencing, the nucleotide sequences generated have a length of 100 bps, even though the length may range from 30 to over 10,000-nucleotide bps. The powerful deep sampling of transcriptomes of very short fragments can be computationally used to reconstruct complete RNA transcripts by using reference genomes for complete alignment (Wang et al., 2009). RNA-Seq has a lot of advantages over microarray due to its five orders of magnitude in transcriptome analysis. Moreover, the quantity of RNA used for analysis expression in RNA Seq is very low as compared to microarray technology (Hashimshony et al., 2012). Theoretically, speaking RNA-Seq has no upper limit for the quantification of RNA transcripts (Ozsolak & Milos, 2011).

The technique is useful for identifying the active genes and their location in the genome and modeling of gene expression in a cell or organism in a specific time period (Tachibana Chris, 2015). By virtue of its large advantages over microarray technology, RNA-Seq has almost dominated almost all transcriptomic techniques until 2015 (Su et al., 2014). It is evident from the existing literature that RNA-Seq is well operated for single-cell transcriptomic data acquisition (Lee et al., 2014).

The RNA-Seq method includes transcript enrichment, fragmentation of transcripts, amplification of transcripts, single-end and paired-end sequencing, and then preservation of data. Further sensitivity of methodology can be increased by enriching the different classes of transcripts and decreasing the known transcripts. It is reported that the fragmentation method is a key aspect for the construction of a library for sequenced transcripts (Knierim et al., 2011). PCR-aided amplification helps to form cDNA copies of transcripts to enrich the fragments for preparation to sequence the expression transcripts (Parekh et al., 2016). During the preparation of libraries the short random sequences called as unique molecular identifiers (UMIs) are individually tagged to the fragments to be sequenced (Kivioja et al., 2011). The UMIs are usually employed for single-cell RNA-Seq transcriptomics, wherein the input RNA is less and amplification is required (Islam et al., 2014). In the preparation of transcripts, the sequencing is done in one (single-end) as well as both directions (paired-end). The application of the strand-specific methods to sequenced transcripts is important for complete sequence information (Levin et al., 2010). Current RNA-Seq method relies on simply copying the transcripts into cDNAs before going for sequencing. Therefore, advancements in RNA-Seq grossly depended on the use of DNA sequencing technologies (Goodwin et al., 2016; Loman et al., 2012).

2.1.3 Massively parallel signature sequencing

Massively parallel signature sequencing (MPSS) is a gene expression array technique that aids to determine the mRNA expression levels by counting individual mRNA molecules, expressed from each of the genes. This MPSS is the newest available technique to conduct expression profiling where there is no requirement of identifying the gene. The sensitivity of MPSS is up to the analysis of a few mRNA molecules per cell, and the obtained data are digitalized, making it simple to analyze the data. MPSS is the most danced and opted technology among both microarray and nonmicroarray-based technologies used to facilitate the complete information of transcriptomes to facilitate strong hypothetical-driven experiments in the current in-silico-based biological era. In 2004, MPSS was used to validate the expression analysis of 104 genes in *Arabidopsis thaliana* (Meyers et al., 2004). About 105 transcripts sequenced by using 454 technology were the earliest published accomplished by using RNA-Seq technology for expression studies (Bainbridge et al., 2006).

2.1.4 Transcriptome-wide association studies

Transcriptome-wide association studies (TWAS) is another gene expression analysis approach, which deals with modeling the regulatory machinery genetically regulated by the expression of genes (Wainberg et al., 2019). This approach critically helps in detecting the potential diseases associated with genes only by using the genetic information of an individual organism under investigation (Gamazon et al., 2015). The phenotype of several diseases such as schizophrenia, Alzheimer's disease, and autoimmune diseases have been accomplished by using TWAS (Gusev et al., 2018; Hu et al., 2019).

3. Metabolomics

Metabolomics is a high-throughput, nonbiased, and comprehensive collection of applied studies employed to resolve (quantify and identify) the complex mixture of total metabolites in a system. Also called as metabonomics or metabolic profiling, metabolomics is the systematic identification, as well as quantification of all the small metabolites synthesized by the body fluids, cells, tissues, organs, and organisms at a particular situation or time. Almost 40,000 metabolites have been identified in by studying human metabolome (Wishart et al., 2012). With studies backed by metabolomics, the concept of systems medicine grossly supports the investigation of patients on discrete physiological, biochemical, and environmental interfaces (Beger et al., 2016; Shah & Newgard, 2015; Sun et al., 2017). The metabolomic studies allow the global assessment of cellular dynamics in relation to environments, considering the level of changes in metabolic reactions, enzymatic kinetics, gene expression, and regulation (Griffin & Shockcor, 2004; Mendes et al., 1992; Mendes et al., 1996). More recently, the metabolome-wide association study (MWAS) is employed, which combines the studying metabolome of samples such as cerebrospinal fluid (CSF) in combination with publicly available data pertaining to genome-wide association study (GWAS) summary statistics (Kim et al., 2021). The major steps in MWAS include the following steps:

- Identification of the single-nucleotide polymorphism-metabolite associations
- Genotype-based building of metabolome prediction models
- Testing of metabolite—phenotype associations with GWAS summary statistics

3.1 Technical intervention

The metabolomics studies are carried out by mass spectrometry and nuclear magnetic resonance (NMR) spectroscopy techniques as depicted in Fig. 10.2. The studies include investigating substrates and their products from cells and tissues. The metabolites ranging from 50 to 500 daltons, including sugars, lipids, amino acids, fatty acids, alkaloids, and phenolic compounds are usually analyzed.

To retain data from the samples, the following steps are important:

- Collection of samples
- Preparation of samples for analysis
- Analysis of samples by appropriate techniques

3.1.1 Collection of samples

For metabolome analysis, both in vitro and in vivo samples can be processed (Chetwynd et al., 2017). The samples may be of diverse origins

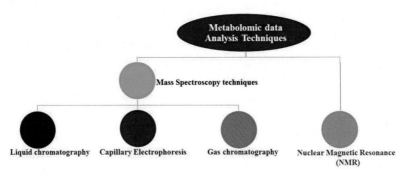

FIGURE 10.2

The flowchart shows the analytical techniques employed for metabolomic studies.

and types such as fluids, cells, and tissues. Among all these samples, fluids are the easiest to acquire and handle for purpose of analysis. Fluid samples include plasma, serum saliva, urine, and fecal fluids. Most of the sample types derived from humans for metabolomic studies include blood or urine samples. Even for nervous and psychiatric disorders, CSF is specially used for carrying out metabolic analysis (Schwarz & Bahn, 2008; Simrén et al., 2019). It must be kept in mind that all these samples must be handled carefully to get desired results, because it is evident that environmental conditions largely influence the metabolic pathways and their products. Proper temperature and steady sample extraction are important for getting desired results.

- **Sample Preparation and biotechniques employed for resolving the compounds in samples:**

♣ Nuclear magnetic resonance:
 To resolve the complexity of samples the techniques such as NMR spectroscopy and mass spectrometry (MS) (Amberg et al., 2017, pp. 229–258). The basic principle of NMR depends upon the ability of spin-active nuclei to absorb and emit electromagnetic radiations when the sample is placed in a magnetic field (Emwas et al., 2015a,b; Markley et al.). The nuclei of hydrogen atoms found in biomolecules are mostly targteted which is most abundantly found in biosamples. The interaction electromagnetic field with the nuclei in the analyte helps to provide information

regarding the chemical environment, motion, and molecular structure. Moreover, it is observed that NMR-based analyses are nondestructive methods of analysis, and the results are highly reproducible for both solid and liquid samples. The NMR technique gives exact information about a wide range of biomolecules and provides in-depth information at the atomic level.

♣ **Mass spectrometry techniques:**
Mass spectrometry is an analytical process in which samples are converted into the gas phase and separated based on the mass/charge ratio. Subsequently, the ions are analyzed by a detector based on a number of ions by mass/charge ratio (Gowda & Djukovic, 2014). The data obtained are compared with the available spectral data from databases to predict the molecular identity of variable constituents. The accuracy and sensitivity of detection are grossly dependent on the setting of the instrument and the standard experimental conditions.

  • **High-performance liquid chromatography:**
  The MS techniques include gas chromatography, which helps to resolve the biomolecules in the gas phase, the majority of these compounds include volatile compounds (Emwas et al., 2015a,b). Further, the separation of compounds is enhanced by using high-performance liquid chromatography (HPLC), which includes chromatographic columns filled with microparticles for withstanding high-pressure elution of samples (Gika et al., 2014).

  • **Capillary electrophoresis:**
  The separation of compounds in samples is based on electrokinetic separation (Maier & Schmitt-Kopplin, 2016). A wide range of compounds can be separated ranging from small inorganic to high MW proteins. Several preanalytical interventions are important, which include processing and addition of analytes.

  ✔ Almost all the techniques mentioned earlier require a very less quantity of samples for analysis.

3.2 Workflow for metabolome analysis (Fig. 10.3)

For data analysis different metabolomic features such as metabolite concentrations, spectral bin areas, and spectral peak areas (Alonso et al., 2015). In addition, univariate and multivariate statistical

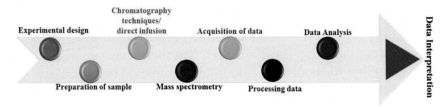

FIGURE 10.3

Workflow of metabolic data from experimental design to data interpretation.

analysis is also performed for higher-order refined results. Multivariate analysis helps to identify the imputed metabolic features, which is least accomplished by univariate analysis. Metabolomics directly links with the phenotype of an organism; hence, it is a versatile technical set of tools important in the field of pharmaceuticals, agricultural sciences, and healthcare sector, apart from its critical application in scientific laboratories for analysis of metabolites. In near future, the personalized metabolome data will enable tracking the precision of drugs and enhanced strategies to design and recommend the drugs on time. Thus, shuttling from population-based drugs to personalized drugs for more effective and precise drug treatment.

3.3 Applications of metabolomics

High-throughput studies pertaining to pharmacology, toxicology, and health ailments regarding nutrition and inborn metabolic disorders can be obtained by metabolomic studies (Spratlin et al., 2009) (Fig. 10.4). Other fields, which need the direct intervention of metabolomic studies, include food technology, microbial biotechnology, enzyme discovery, plant and animal biotechnology, and systems biology. Several applications such as characterization of diseases such as wide range of cancers, autoimmune diseases, diabetes, Alzheimer disease, atherosclerosis, and coronary diseases can be precisely diagnosed and assessed by applications of metabolomics studies. The status of the metabolome in an organism could help to predict the phenotype based on the insertion or deletion of a gene. Some of the major applications of metabolomics are briefly explained in the following sections.

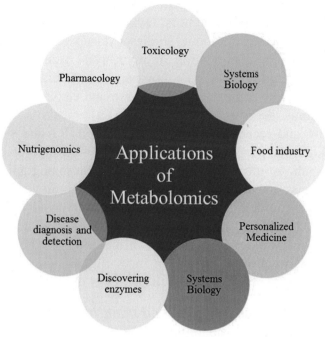

FIGURE 10.4

Demonstration of metabolomics applications in the field of human health, disease biology, plant, and environmental biology.

3.3.1 In human health and diseases

Oncological studies of human beings can be deeply studied by employing metabolomics. Cancer cells need to be evaluated thoroughly due to their differential metabolic requirements such as high-energy investment, through unregulated gene expression and metabolism (Hanahan & Weinberg, 2000). The metabolomic studies resulted in the identification of cancer biomarkers during preclinical studies during preclinical treatments and for validation of these biomarkers in biofluids such as prostate cancer fluids, urine, blood, and other fluids. In combination with genomics and proteomics studies, metabolomics helps in precise diagnoses and prevention of a wide range of cancers.

In addition to cytokine markers, several protein markers such as differential expressed macrophage-stimulating proteins and macrophage-derived chemokines have been identified in liver carcinomas (Liu et al., 2011). In certain tumors such as neuroepithelial tumors, gliomas, and brain tumors, enhanced levels of amino acids were identified by metabolomic studies (Masuo et al., 2009). The metabolomic-based identification of myocardial ischemic episodes is critical to diagnose and cure coronary artery diseases (Griffin et al., 2011). In this context, several protein and enzyme biomarkers, such as tumor necrosis factor α (TNF-α), reactive C-protein, interleukin-6 and receptors types 1 and 2 (sTNF-R1 and sTNF-R2) used for diagnostic purpose (Pai et al., 2004). Moreover, metabolic approaches such as ultra-performance liquid chromatography—high-resolution mass spectrometry technology were employed to screen the biomarkers for identification of polycystic ovary syndrome women (Yu et al., 2021). Identification of urinary biomarkers in acute kidney injuries by metabolomic profiles of urine after cardiopulmonary bypass surgery by ultra-performance liquid chromatography (UPLC)/MS is a promising approach (Beger et al., 2008).

3.3.2 In agricultural sciences and plant biology

The identification as well quantification of plant metabolites is critical to understand the metabolite dynamics and their flux in metabolic pathways and physiological stimuli (Fiehn & Weckwerth, 2003). About 200,000 metabolites have been identified in plants to play a critical role in physiology and adaptation in diverse ecological niches (Fiehn, 2002; Paine et al., 2005). Therefore, it is evident from the huge metabolome data that plant metabolites are too complex and diverse and need critical assessment for analyzing data sets at pM to mM concentrations. We now believe that plant metabolites play a critical role in defining the phenotype due to their role in contributing to the taste, color, aroma, and physiological attributes of fruits and flowers (Bino et al., 2004). Plant metabolomics studies play a critical role in deciphering the metabolic a biochemical network responsible for the developmental and physiological role of cells, tissues, and organisms (Bino et al., 2004; Weckwerth, 2003). The improvement of genetically modified organisms is accomplished through in-depth knowledge of metabolites through metabolic studies. The improvements through the intervention of metabolomic help to estimate the risks associated with the pesticides resulting in having information regarding the complexity of biochemical and physiological implications of chemicals with crop plants. In addition, plant metabolomics is also used to assess the changing metabolite dynamics during stress conditions such as cold stress (Cook et al., 2004). The critical applications of plant metabolomic studies include gene function discovery, engineering the metabolomic pathways to biochemical and pharmaceutical pathways (Bhalla et al., 2005). The combination of transcriptional profiles with metabolomic profiles can provide a platform to identify the functions of unknown genes of both model plants as well as nonmodel plants (Saito et al., 2008).

3.3.3 Personalized medicine
Personalized medicine is an applied branch of medical sciences that uses the individual genome, that is, genetic profile to cure patients through prevention, diagnosis, and disease treatment. Medical practitioners provide medication, therapy, and administration to a patient based on their genetic profile. In recent times, metabolomics has further augmented personalized medicine for quick medical diagnosis to identify diseases. Metabolomics offers a rapid and in-depth profile of metabolites found in diseased individuals through untargeted and targeted approaches. Being highly suitable and informative, metabolomics is promising in aiding pediatric medicine (Carraro et al., 2009).

3.3.4 Metabolomics in aid of nutrigenomics
The field of nutrigenomics combines the application of transcriptomics and proteomics in addition to genomics and metabolomics to study the effect of daily nutrition on the genome and its dynamics. Scientists strongly believe that the metabolome of an organism strongly depends on the intake of nutrients from food and medications apart from gender, genetic susceptibilities, age factor, and other concurrent health conditions. The application of metabolomics to nutrigenomics is used to assess the metabolomic fingerprint meditated by exogenous and endogenous factors on the body metabolism of an individual.

3.3.5 Metabolomics aiding in detection of changes induced by environmental factors
The effect of the environment on organisms is mediated by several biotic and abiotic factors, which in turn continuously change the metabolites of organisms. Apart from other omics technologies,

metabolomics is at the forefront to assess the health of an organism in the changing environment based on the metabolomic profiles.

4. Interactomics

In the field of applied sciences, which deals with studying the protein—protein interaction to untangle the regulation and functions of a diverse range of total proteins in a cell or organism, the studies of interactomics aid in deciphering the causative mechanism of diverse diseases, discovery of drug targets, and development of novel drugs. The field of interactomics is the fusion of biochemistry with engineering and informatics to provide insights into the global overview of protein—protein interactions and their networks and the workflow employed for studying is demonstrated in Fig. 10.5. The critical aim of interactomics is to understand the pathophysiology of various diseases, define novel biomarkers, and develop new drugs based on interactions and docking. Several approaches have been at the forefront to study the interactome of an organism, which include in vitro, in vivo, and in silico (Rao et al., 2014) (Fig. 10.6).

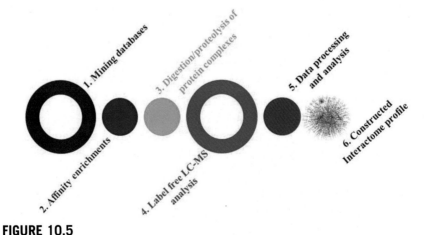

FIGURE 10.5

Schematic interactomics workflow employed to accomplish the PPIs by employing several techniques and software to uncover interactomes.

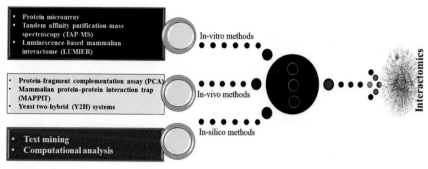

FIGURE 10.6

Schematics to represent the technical machinery employed for studying the interactome of a cell/organism.

4.1 Methodology of interactomics

4.1.1 The in vitro methods for interactomics

The in vitro method is based on performing experiments in controlled laboratories but outside the living organism. The methods employed for studying interactome in in vitro conditions include protein microarray, tandem affinity purification-mass spectrometry (TAP-MS), and luminescence-based mammalian interactome (LUMIER) technique. The TAP-MS is evolved to overcome the limitations of the AP-MS technique to eliminate the high false positive rates (Rigaut et al., 1999). The target protein is tagged to a protein composed of three major tags, like protein A having a high affinity with IgG immunoglobulin, and another peptide (calmodulin binding peptide) having a high affinity with calmodulin. Both of these tags are separated by cleavage sites recognized by the protease encoded by the tobacco etch virus (Burckstummer et al., 2006). The target protein tagged with TAP is first purified by using protein-A affinity support and then subsequently cleavage by incubating with tobacco etch virus protease to release the target protein. In another affinity step, the whole protein complex is purified by tagging with calmodulin-binding peptide in the presence of calcium and subsequent release by employing calcium-chelating chemicals such as ethylene glycol tetraacetic acid (Figeys, 2008). The purified protein complexes are further subjected to mass spectrometry analysis. The MS analysis provides specific and sensitive identification of proteins and the advantage to examine the multiple interactions between multiple proteins (Porras et al., 2012).

LUMIER is the combinatorial approach of biochemical strategies and a two-hybrid system to have a high throughput PPI analysis in cultured cells (Barrios-Rodiles et al., 2005). In this technique, the prey protein is attached to the Renilla luciferase enzyme (for detection) and the bait protein is attached to the affinity tags such as FLAG (for purification). The weak and transient interactions between overexpressing proteins may yield false positive results during LUMIER analysis of PPIs (Petschnigg et al., 2011).

4.1.2 The in vivo methods of interactome

As far as in vivo methods of interactome are concerned, they include performing experimentation within a living organism by the intervention of techniques such as protein-fragment complementation assay (PCA), mammalian protein–protein interaction trap (MAPPIT), and most notably yeast 2-hybrid (Y2H) systems. All the methods operate through the common principle of molecular fishing and require the interaction of bait and prey proteins (Ivanov et al., 2011). The PCA is based on the reconstitution of reporter protein from nonfunctional fragments by direct physical interaction. The techniques rely on refolding of the protein fragments and do not depend on the localization of the protein (Stynen et al., 2012). The PCAs are also known as "split-protein sensors" and specifically they can be named on the basis of protein under study like split-X methods, for example, split-FP, split-DHFR, etc. (Stynen et al., 2012).

The second method, MAPPIT, is another potential in vivo method to study the PPIs and is based on the reconstitution of signal transduction pathways initiated by the interaction of bait and prey. The interaction of bait (ligand of receptor) and prey (receptor) drives the signal pathway to say for example activation of MAPKKK pathways through STAT-3 (Eyckerman et al., 2001). Limitations in previous MAPPIT methods are overcome by new derivative MAPPIT such as Kinase Substrate Sensor, in which the bait and prey can be either membrane bound or soluble to apply the technique to a wide range of proteins (Lievens et al., 2014).

Fields and Song (1989) proposed that the Y2H systems involve the complementation of two proteins, viz, bait and prey (Fields & Song, 1989; Porras et al., 2012). The Y2H usually consists of two domains with special functions such as a domain of transcription factor (bait) and interaction partner protein (prey). Both the proteins upon interaction lead to the formation of a complete transcription factor and hence the expression of reporter genes as demonstrated in case of yeast (Stynen et al., 2012).

4.1.3 The in silico interactome analysis

In in silico interactome analysis, the text mining and then computational analysis are carried out by computational simulations. Text mining involves the collection of data pertaining to interactome obtained through studies based on low-throughput protein interaction (Lievens et al., 2010). Text mining can be done either manually or by the application of different semantic algorithms (Zhou & He, 2008). The publicly available databases to fetch the interactome-based text include the Human Protein Reference Database (Keshava Prasad et al., 2009), Munich Information Center for Protein Sequence (Pagel et al., 2005), the Molecular Interaction database (Chatr-aryamontri et al., 2007), and many more.

The experimentally determined is further supplemented by a variety of computational methods such as the structure-based approach, the coevolution analysis approach, and the ortholog-based approach. The structure-based method can predict the protein—protein interactions in case the two proteins are similar in structure (Ogmen et al., 2005). Once 3D structure of a protein is predicted, it is easy to predict the interacting domains of similar structured proteins. The assessment of structural prediction methods can be done by suing the critical assessment of structure prediction (Moult et al., 2011). As far as, the ortholog approach is concerned only orthologs of gene and their organization in genomes are constructed to correlate their functional relationships (Ivanov et al., 2011). In this method, the PPIs are predicted on the basis of evolutionary relationships and by using phylogenetic profiles to predict the functional relationships of proteins (Fang et al., 2010). Due to functional linkage in genomes, proteins have strong pressure to be inherited together due to the process of evolution (Shi et al., 2005). Limitations of this method lie in unexpected disturbances due to duplications that may deviate the predictions based on the phylogenetic profiles and it has been reported that this method is more reliable in predicting protein—protein interactions in prokaryotes rather than eukaryotes (Lin et al., 2013).

4.2 Applications of interactomics

Interactomics plays a critical role in investigating a wide range of applications such as in unraveling the cellular and molecular dynamics of cells, to study diseases, discovery of biomarkers, etc. A few important applications of interactomics are explained later and demonstrated in Fig. 10.7.

4.2.1 In investigation of cellular and disease biology

The key regulatory networks linked to the health and disease related to proteins can be untangled by studying the PPIs through interatomic studies. For example, pyruvate kinase isozyme M1/M2 (PKM2), a novel regulator of D23—230 cellular prion protein (PrPC), was identified in neuronal cell lines of the murine hippocampus by suing chromatography/MS analysis by interatomic approach (Zafar et al., 2014). Several signaling pathways, such as ERK signaling pathways, have been addressed by employing interatomic studies (Kolch & Pitt, 2010). The diverse diseases such as cardiomyopathies,

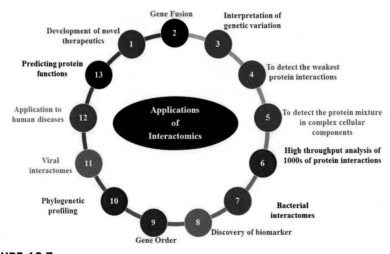

FIGURE 10.7

Wide range of applications accomplished by the interatomic studies of cells/organisms.

lipodystrophies, muscular dystrophies, and premature aging syndromes caused by mutations in lamin A gene were detected by Y2H screening is another classical example of interatomic studies (Dittmer et al., 2014).

4.2.2 Discovery of biomarker

Interactomics aids in the discovery of biomarkers to provide functional details regarding the role of proteins and their associated genes. Interactomics has been used to identify the potential biomarkers for Ewings sarcoma, lung cancer, and breast cancers (Gamez-Pozo et al., 2012; Haga et al., 2013; Korwar et al., 2013).

4.2.3 In uncovering the drug targets

It is estimated that the total number of human PPIs ranges between 130,000 and 650,000, most of which are related to disease and can be potential targets of a large diversity of drugs (Stumpf et al., 2008; Venkatesan et al., 2009). Drug discovery through the application of interactomics can be further used in unraveling pathogen—host interactions. A combinatorial approach of interactomics with protein—DNA interactions, expression levels, and phenotypic data can further augment the discovery of newer drugs for a wide range of diseases.

5. Lipidomics

The field of omics technology deals with structural-functional aspects of the complete lipid profile of lipids produced by a cell or an organism. In addition, lipidomics also helps to decipher the interaction of lipids with other lipids, proteins, and a large number of metabolites. It is critical to study the lipids and their dynamics, because lipids are important for cellular functions such as membrane formation and biosignaling and act as energy repositories. Lipidomics emerged in 2003 and thereafter it is greatly explored by biotechniques such as mass spectrometry (MS) and chromatography techniques, such as thin layer chromatography, gas chromatography, or HPLC (Fig. 10.8). MS in combination with chromatographic techniques helps in the separation and detection of lipids with structurally similar belonging to the same class. In all, two major analyses based on these

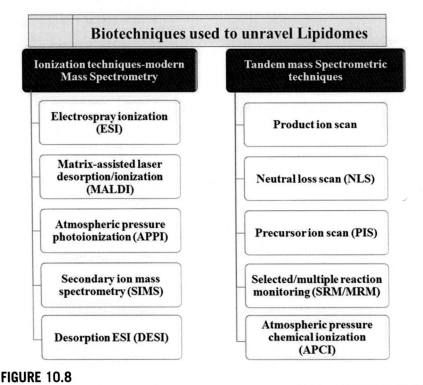

FIGURE 10.8

List of biotechniques employed for analyzing the lipidome of a cell or organism.

technical methods, viz., Global lipid analysis profiling and targeted lipidome analysis can be carried out to unravel the lipidomes of specific cells or organisms.

5.1 Technical intervention

5.1.1 Mass spectrometry for studying lipidomes

The studies of lipidomes involve sampling, storage, and then processing for analysis by appropriate lipidome techniques. The classical workflow used by the mass spectrometry technique is shown in Fig. 10.9. It is pertinent to consider factors such as condition of sample, preprocessing of samples, and proper selection of the type of study when proceeding with sampling. Preparation of samples is done as per the technique to be applied for lipid profile analysis. Nonextracted

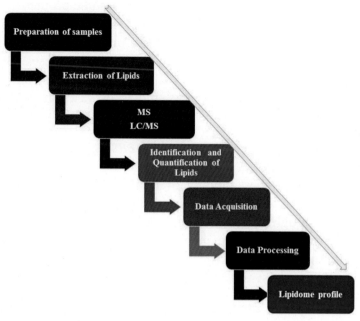

FIGURE 10.9

Workflow for generating lipidomic profile of a cell/organism under investigation.

tissue slices are analyzed by using MS imaging. For quantitative lipidome analysis, it is critical to add appropriate internal standards such as normalization to total proteins, fluid volume, or dry to wet weight of tissue. Of shotgun lipidomics and chromatography-based lipidomics, in particular, liquid chromatography-based Lipidomics is used for MS analysis of lipidomes (Fauland et al., 2011; Han et al., 2012).

5.2 Applications of lipidomics

5.2.1 To investigate metabolic syndromes

Metabolic diseases such as diabetes, cardiovascular diseases, strokes, and fatty liver (nonalcoholic) are debilitating lipid-linked complications found in humans. These diseases are further increased due to the rise in humans with obesity. In this context, current advancements in lipidomics studies play a key role in the timely prediction and quantification of risk and in paving way for therapeutic solutions to these metabolic syndromes through the studies of lipidome profiles. Lipidomics aids in detecting the lipoproteins and plasma lipid fractions to provide insights regarding the high-density lipoprotein lipidome and hence their adverse effects to aggravate vascular diseases. Studies such as pathogenesis, population profiling, and biomarker identification are grossly aided by lipidomic studies of vascular health and ischemic heart diseases.

5.2.2 In studying the neurological diseases

Most neurological disorders are associated with lipids because the brain harbors the highest amount of lipids. The lipids play a vital role in homeostasis, metabolism, signaling, and trafficking. Valuable information through the studies of lipidomes has helped to assess the health status of the human brain. Scientists have studied biomarkers for tracking the early diagnosis and prognosis of serious neurological disorders.

5.2.3 To study and investigate the cancers

The key role in cancer development is identified to be played by lipid molecules because we know that lipids play an important role in the growth and metabolism of cells and organisms. In particular, nonesterified fatty acids play a critical role in the building and biosynthesis of phospholipids, sphingolipids, and cholesterol. The latter class of lipids are major components of the plasma membrane and act as reservoirs of energy for cells. Bioactive lipids such as lysophospholipids play an important role in biosignaling and in the production of second messengers, which in turn may aid in cancer development, its survival, and migration in unregulated cellular conditions. Likewise, cleavage products of phosphatidylinositol such as diacylglycerol and isopentyl triphosphate (IP3) are pivotal second messengers for activating different signaling pathways such as PI3K/AKT signaling pathways important for the development of cancers. The change in the dynamics of these critical lipids can be grossly asses by lipidomic studies. Later studies provide a golden window to explore, diagnose, and aid therapeutics for a wide range of cancers. Through the lipidome of blood and other fluids, one can identify and diagnose cancers.

5.2.4 In detection of eye diseases

To diagnose and understand eye disease, it is reported that lipidomics plays a pivotal role. The ocular surface disease and health of tear film are deeply investigated by lipidomic studies (Pieragostino et al., 2015). The replenishing of lipid tears can be further effectively cured by the application of lipidomic studies to reduce the inflammation and clinical features of the disease (Zhou & Beuerman, 2012).

5.2.5 In studying the nutrition

The lipid content in daily nutrition can be studied by using lipidomics to investigate the composition, structure, and functional aspects of lipids (Smilowitz et al., 2013). The functionality of lipids in nutrition such as their existence as signaling molecules, metabolic intermediates, and nutrient sensors is of prime importance to enhance the quality of nutritional profiles. By investigating the lipids in nutrition, one can assess their chromic, medium, and acute effect on dietary products. The status of health-benefiting fatty acids such as ω-3 polyunsaturated fatty acids can be assessed by carrying the lipidome studies in dietary nutrition (Maskrey et al., 2013; Smilowitz et al., 2013).

5.2.6 In screening and discovery of drug and screens

The investigation of screening of drugs, toxicity evaluation, preclinical testing, and more recently development of personalized is accomplished by the application of lipidomics. Exploration of lipogenesis inhibitors by lipidomic studies helps to circumvent the designing of drugs against a wide range of cancers and biomarker identification.

6. Metagenomics

The field of metagenomics was coined to represent the field of omics that deals with investigating the genomes of microorganisms for unison identification in habitats like soil and water. The field involves the isolation of total genomic DNA from a particular environment and then the assessment of species diversity based on species-specific DNA sequences. The field emerged with the aim to have insights regarding genetic diversity and population structure and most especially to diagnose the species diversity found to be associated with different organisms in their organs. Before the last 3 decades, Pace et al. (1986) revolutionized the idea that DNA sequences can be directly cloned from different habitats to analyze samples for the microbial populations to reveal the complexity of microbiomes. Later on, the term "metagenome" was used by Handelsman et al. (1998) to highlight the importance of the field in identifying the soil microorganisms directly from the natural source/habitat.

Metagenomics deals with the identification of microorganisms, which are nonculturable. Hence, species-specific markers such as 16S ribosomal RNA sequences are utilized for the identification of individual species among the samples. The 16S rRNA sequences are conserved and relatively short sequences, which differ from one species to species. Several applied studies direly need to obtain pure line cultures for surveillance and response of microorganism of interest is at helm. At the crossroads of this investigation, metagenomics was a major breakthrough to identify the microbiomes in major habitats of living organisms. It is a powerful tool for redefining the hypothesis of microbial functions, which is evident from the identification and in fact discovery of ammonia-oxidizing Archaea or proteorhodopsin-based photoheterotrophy (Beja et al., 2000; Nicol & Schleper, 2006). The current application of metagenomics deals with targeting the identification and explorations of local microbial communities associated with the specific environment. Sanger sequencing technology provided pivotal progress in the field of metagenomics during the initial era (Breitbart et al., 2003; Gillespie et al., 2002; Sanger et al., 1977; Uchiyama et al., 2005). Even though later nowadays, the field of metagenomics is revolutionized by the intervention of NGS technologies (Klindworth et al., 2013; Oulas et al., 2015; Sunagawa et al., 2015). As reported, it is found that 10% of bacterial or archaebacterial sequences were not recovered by using 16srRNA/SSU RNA (Eloe-Fadrosh et al., 2016). The problem of incomplete identification was addressed by elucidation of bacterial phylogeny by using appropriate markers such as elongation factor Tu (EF-Tu) and heat shock protein 70 (HSP70) (Venter et al., 2004).

6.1 Methodology employed for metagenomics

The field of metagenomics is divided into the following two main approaches:

- **Structural metagenomic approach**
- **Functional metagenomic approach**

6.1.1 Structural metagenomic approach

This approach of metagenomics mainly focuses on studying the uncultivated microbial populations from the natural habitat, because these studies may help to reconstruct the metabolic networks among diverse members of the microbial community members (Handelsman, 2005; Tringe et al., 2005). Studying the community structure provides an in-depth understanding of the relationship between different individuals of the community with other members to build an essential component of the ecosystem and unravel the ecological as well biological functionalities of all the members of the community (Tringe et al., 2005; Vieites et al., 2009). Even though most of the 16SrRNA-based taxonomic identifications are referred to as metagenomics, this is not the case in real. For example, structural genomics aims to investigate the microbial communities and allows reconstruction of the community structure to reveal the metabolic profile of microbiomes to validate the role of microbial communities (Guazzaroni et al., 2009; Handelsman, 2005; Louca et al., 2018; Tringe et al., 2005).

6.1.2 Functional metagenomic approach

This approach aims to identify the genes coding for proteins or RNA molecules. The approach is based on the production of expression libraries with almost thousands of metagenomic clones followed by screening based on activity (Guazzaroni et al., 2015; Schmeisser et al., 2007).

The general approach of metagenomics involves studying uncultivated microorganisms by suing a whole genome shotgun. Further DNA samples are either fragmented by shearing also called as shotgun method or sequenced by suing NGS or Sanger sequencing method (Fig. 10.10). The sequenced bacterial genomes are then characterized on the basis of 16SrRNA (Tringe et al., 2005). Metagenomics clearly sheds light on a field like evolutionary biology to understand the constraints in variation of the genome through the understanding of sexual reproduction (Gorelick & Heng, 2011). Moreover, the metagenomic approach helped to reveal that microbial species are not clonal in nature (Venter et al., 2004).

6.2 Applications of metagenomics

A large number of applications have been reported with respect to metagenomics in the field of evolutionary biology, microbial ecology,

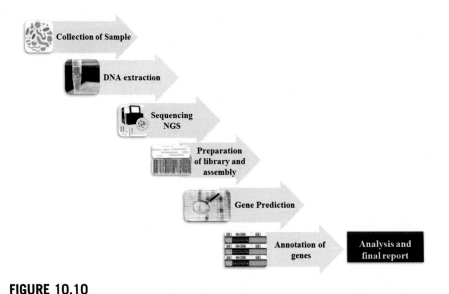

FIGURE 10.10

Technical and analytical schematics of metagenomic process.

human health, etc. A few of the critical applications of metagenomics are precisely mentioned in the proceeding sections and demonstrated in Fig. 10.11.

6.2.1 Unraveling the genomes and reconstruction of phylogeny

The taxonomy of microorganisms has been updated by covering the uncultivable microbial species, which were not identified by other methods of taxonomy. The knowledge of genomes from Sargasso Sea waters by shotgun sequencing is the best example to display the potential of metagenomics as a feasible technique for the exploration of microbial diversity (Venter et al., 2004).

6.2.2 Reconstruction of microbial communities

Metagenomics helps to reconstruct unexplored microbial species. For example, the assembly of genomes from Leptospirillum group II and Ferroplasma type II was directly obtained from environmental samples (Tyson et al., 2004). The comparison of functionality and composition of microbial communities has been studied in nutrient-poor and nutrient-rich habitats by Tringe et al. (2005). The approach led to decipher the gene functions in the target genomes.

FIGURE 10.11

Pictorial representation of diverse metagenomics applications in human health and disease, agricultural and environmental sciences.

6.2.3 A tool to untangle the microbiomes from humans

The last few decades have led to unravel the microbiomes associated with humans and their role in controlling different physiological and metabolic functions (Archie & Tung, 2015; Gilbert et al., 2016; Kau et al., 2011; Lloyd-Price et al., 2017). By carrying out metagenomic studies in different features of New York subways about 1700 microbial taxa were found to be dominating the human skins rather than urogenital tracts and GI tracts (Afshinnekoo et al., 2015). It is estimated that the human gut harbors about 10^{14} microorganisms, which include bacteria fungi, viruses, and protozoa (Arnold et al., 2016; Gill et al., 2006).

6.2.4 In human health

The metagenomics studies through the use of NGS technologies help in the production of bacterial genomic profiles to study their relation with genetic variation and microbial diversity and their impact on positively or negatively regulating human diseases. Human is a super-organism, in the sense it harbors some 10^{13} microorganisms with multiple roles in the regulation and generation of diseases (Bäckhed et al., 2005).

6.2.5 A tool to aid the bioremediation process

Metagenomics is a potential tool to improve the strategies used for monitoring the pollutants in different ecosystems and it helps to decontaminate the different habitats. The field has been at the forefront to unravel the microbial communities associated with the biaugumentaion and bioremediation. Due to its direct investigation approach, metagenomics aids in the exploration of vast microbial diversity that is responsible for activating the degradative pathways important for the bioremediation process.

6.2.6 Industrial applications of metagenomics

The uncultivated microbial species are of prime importance to the majority of industries because they play a critical role as biocatalysts, and as a repository of elusive metabolites. Metagenomics helps to discover novel bioactive compounds and biocatalysts of industrial importance (Daniel, 2004; Osburne et al., 2000). Together with high throughput and in vitro evolution, techniques of metagenomics provide a great opportunity for industries to unravel the repository of active metabolites having potential applications in bio-industries.

6.2.7 Advancements in microbial ecology

A massive amount of data pertaining to the microbiomes and their functional diversity helps to reconstruct the phylogeny based on metagenomic studies. For example, this approach has been used to reconstruct the metagenome of acidophilic biofilms having very low diversity (Tyson et al., 2004). Metagenomics has evolved as the best promising tool to investigate microbial ecology, evolution, and more specifically diversity in extremely sensitive habitats.

7. Ionomics

The field of ionomics deals with the assessment of the complete set of mineral nutrients and trace elements in a cell organism, in particular plants. The ionone is actually the inorganic subset of metabolites belonging to a metabolome of a cell/organism and was originally used for the identification of global metabolomes of *S. cerevisiae* and *Escherichia coli* (Oliver et al., 1998; Tweeddale et al., 1998). Ionomics also deals with the investigation of molecular mechanisms underlaying the compositions of elements in an organism. Such elucidation of biochemical pathways is used to unravel the production of ions critical for integrity of cells, importance in enzymes as cofactors and transportations systems (Zargar et al., 2016, pp. 317–344). The last 1 decade has witnessed great progress in ionomics since its beginning in 2003 (Lahner et al., 2003) due to the introduction of high throughput techniques to study the effect of ionome on genotype and phenotype (Danku et al., 2013). Technical approaches such as DNA microarray and deletion mapping-based bulk segregant analysis (BSA have greatly accelerated the progress in the field of ionomics (e.g., Baxter et al., 2009; Chao et al., 2011). It is thus evident that ionomics helps in understanding gene regulation in an effective and critical manner. Several plants species have been subjected to ionome analysis such as *Hordeum vulgare* (Wu et al., 2013), *O. sativa* (Norton et al., 2010; Zhang et al., 2014), *Z. mays* (Gu et al., 2015), *Lonicera japonica* (Chen et al., 2009a,b) *S. lycopersicum* (Sanchez-Rodriguez et al., 2010), and *G. max* (Ziegler et al., 2013).

7.1 Technical intervention for studying ionomics

On the basis of ionome profile, the mutants and natural alleles can be differentiated considerably (Lahner et al., 2003; Chen et al., 2009a,b). The alteration of ionome profile has been linked to genes, such as genes ESB1 and TSC10a, needed for control of the deposition of suberin in root endodermis (Baxter et al., 2009; Chao et al., 2011). A diverse range of techniques have been utilized to investigate the ionome of the plants; for example, in *A. thaliana*, inductively coupled plasma mass spectrometry (ICP-MS) helps to decipher the high-throughput elemental profile of ionome (Lahner et al., 2003) (Fig. 10.12). Another technique, ion coupled plasma mass spectrometry analysis, was employed to identify 20 elements among 947 lines of plants.

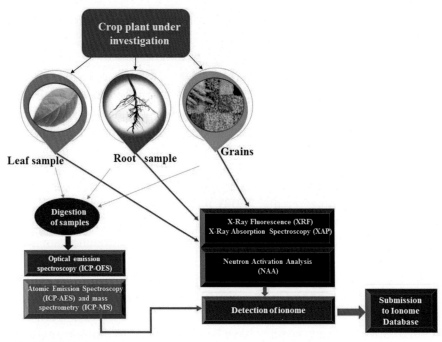

FIGURE 10.12

Representation of the technical approaches and workflow operated for ionome analysis of plant samples.

7.1.1 The inductively coupled plasma (ICP) through the use of optical emission spectroscopy (ICP-OES) or atomic emission spectroscopy (ICP-AES) and mass spectrometry (ICP-MS)

In ICP, the analyte atoms are ionized into gaseous or plasma form to be detected by ICP and ICP-AES or ICP-MS. For generation of plasma, a silica torch is employed located within the argon-cooled coil or water-cooled coil. Subsequently, the plasma gas is introduced into the plasma torch and hence radio frequency ionizes the gas to make it electrically conductive. The flow of argon gas cooling maintains the position and insulation of plasma gas in the instrument. The elements to be analyzed must be in solution. The particulate formation must be avoided such that clogging is avoided within the instrument. Moreover, the aerosol formation from aqueous samples is mediated by the nebulizer. The aerosol form of the sample now passes into the spray chamber in combination with argon gas, the finest sample droplets are swept with plasma, and the rest large sample droplets are settled out and wasted. The ionized analytes are detected by employing either a mass spectrometer or an optical emission spectrometer.

The high throughput ionome profiling has led to the development of databases like iHUB, which can be accessed through www.ionomicshub.org/. One can retrieve and analyze data by using appropriate software application for comparative accounts and updating databases for unraveling the role of ionome in plant development and growth. The iHUB helps to manage the databases retrieved from different research inputs such as information relating to cultivation, harvesting, sample preparation, elemental analysis, and data processing with respect to ionomics.

7.2 Applications of ionomics

Large studies have reported the role of ionome in adapting to a common environment (Anderson et al., 2011). The identification of a large number of natural alleles controlling the variation in ionomic traits has opened a new window of understanding the dynamics of molecular mechanisms of adaptation to diverse environmental conditions. A large number of ionome-controlling alleles, such as HMA3 for Cd, or HAC1 for As, APR2, and ATPS1 for S, have been identified to enhance the adaptability to local environmental conditions. Ionomics possess a large number of applications in living organism and environment (Fig. 10.13).

8. Connectomics

The last few decades have witnessed a rapid pace of advancements in the studied detailed architecture of the brain through the studies of neural connectivity studies of the brain to have deep insights regarding the functional integration and structural details of the brain. The central phenomenon to develop comprehensive maps to develop the map of connections between neurons is studied under the field on connectomes (Sporns et al., 2005). The field deals with the studies and production of the connectomes to unravel the connections established

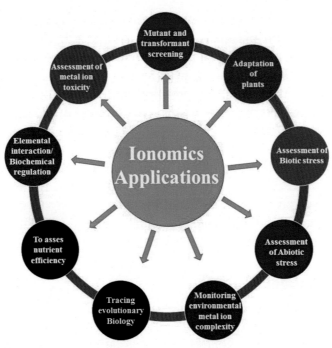

FIGURE 10.13

Representation of an overview of ionomics applications in living organism and environment.

by the nervous system of the organism throughout the body. The goal is achieved through big data analysis to develop structural and functional imaging of the brain. The studies of connectomics are too complex to investigate, so neural scientists employ high throughput techniques such as those dealing with neural imaging and histo-cytological techniques with aim of increasing the pace of resolution and efficacy. It is evident from the existing literature that connectomics deals with fetching the comprehensive mapping of brain connectivity across all dimensions from microscale to macroscale analysis of connections established by neurons (Sporns et al., 2005).

8.1 Technical inputs for connectomics

The wide range of organisms such as fruit fly *Drosophila melanogaster* (Chiang et al., 2011), nematode worm *Caenorhabditis elegans* (White et al., 1986), human (Hagmann et al., 2007; Iturria-Medina et al., 2007; Zalesky & Fornito, 2009), pigeon (Shanahan et al., 2013), macaque (Modha & Singh, 2010; Stephan et al., 2001), mouse (Bota et al., 2012), and cat (Young et al., 1994) are used to study and develop connectomic maps of the brain. The mainstream techniques employed for studying connectomics involve diffusion magnetic resonance imaging (MRI) and 3D electron microscopy with the involvement of fluorescent dyes. The approach may involve the injection of tracers into the candidate brain and their diffusion in brain cells and flushing of tracers to axons of neurons. Then posthistological staining techniques are employed to unravel the connections of brain cells. The tracer techniques being accurate and precise are widely used to develop the connectomes of an organism. On large, the critical to the development of brain connectomes is to investigate the macroaxonal structure, that is, the physical wiring of neurons by employing diffusion-weighted imaging (DWI) and to study the functional dynamic properties adapted by these connectomes by use of functional MRI (fMRI) (Bullmore & Sporns, 2009; Fornito et al., 2013).

8.1.1 *Magnetic resonance imaging*

The use of MRI for mapping of connectomes involves the following major steps:

- ✔ Defining regions
- ✔ Measuring connectivity between these regions
- ✔ Network analysis

A large number of heuristic methods have been employed to diagnose the deceptive boundaries or defining regions of the brain as mentioned in the following bullet points:

1. Priori-anatomical templates (Desikan et al., 2006)
2. Predetermined functional criteria (Dosenbach et al., 2010)
3. Random parcellations of varying resolution (Fornito & Bullmore, 2010)
4. Data-driven parcellations (Yeo et al., 2011)
5. Coregistration with histological data (Eickhoff et al., 2005)
6. Voxel-wise mapping (van den Heuvel et al., 2008)

7. Quantitative mapping of regional variations in specific imaging signals (Glasser & Van Essen, 2011)

The previous heuristic methods have their own characteristics and limitations and may be applied as per the need of results and regions of the nervous system. The measuring of connectivity between the defined regions such as structural, that is, physical wiring of the brain (anatomical connections between neural elements such as axonal, dendritic, and synaptic), functional (statistical dependence between the neurophysiological signals), and effective regions is accomplished by DWI connectomics analysis. The DWI is employed to tractographically reconstruct the putative fiber pathways to link discrete regions in pairs. Whereas, the effective and functional connectivity is usually measured by using fMRI and electo- and/or magneto-encephalography (EEG and MEG, respectively) to get spatial-temporal dynamics of connectomes resolution. By deciphering all the connectivity between the regions, a connectivity matrix is developed, each connectome has an intrinsically directed matrix and an intrinsically weighted matrix (Fornito et al., 2013). Consequently, the connectional architecture of the brain is grossly described by the connectivity matrix and is critical to examine the topology or network connectivity (Fornito et al., 2013).

Topological or network connectivity is important for the arrangement of connections with respect to each other and is developed by suing mathematical studies in the form of graph theory (Bullmore & Sporns, 2009). To assess the healthy state or disease state of the brain, one has to proceed with connectome analysis through the investigation of candidate circuit, connectome-wide, and topological studies.

8.2 Applications of connectomics (Fig. 10.14)

- The studies help to elucidate the computational models pertaining to the whole dynamics of the brain.
- The development of connectograms, that is, the circular diagram of connectomics helps asses the injuries to neural networks to analyze brain injuries.
- Connectomics helps to understand the assessment and diagnosis of neurological disorders such as Alzheimer's disease and

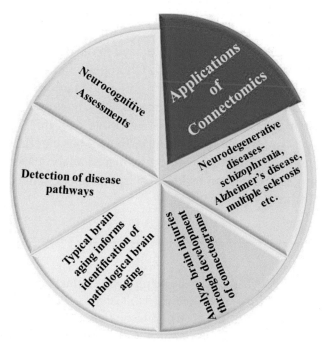

FIGURE 10.14

Representation of critical functions of connectomics in the field of neurobiology.

schizophrenia. So connectomics has emerged as a new field in detecting neurological disorders in a precise and accurate way (Fornito & Bullmore, 2014). The connectomics studies reported that almost all neurodegenerative diseases such as schizophrenia, Alzheimer's disease, multiple sclerosis, and mild cognitive impairment alter the modularity of the human brain (Griffa et al., 2013).

9. Conclusion

Omics is a collective term, which includes high throughput analytical techniques such as genomics, transcriptomics, proteomics, lipidomics, and metabolomics integrated with computational and bioinformatics tools to assess total information regarding genomes, transcriptomes, proteomes, lipidomes, metabolomes, ionomes, and metabolomes.

Several aspects of living systems can be assessed in an unbiased manner by the versatile applications of omics technologies. For example, the development of biomarkers led to the design and delivery of a specific target, thus enhancing the therapeutic potential of drugs and disease management. Moreover, omics technologies have provided deep insights into virulence, stress biology, bacterial physiology, and mechanism of action mediated by drugs (antimicrobial compounds) (Roemer & Boone, 2013; Tang, 2015). Thus careful designing of experiments in conjunction with omics-based analytical techniques and use of software tools can tackle wide range of challenges posed to humans and other living creatures.

References

Afshinnekoo, E., Meydan, C., Chowdhury, S., Jaroudi, D., Boyer, C., Bernstein, N., Maritz, J. M., Reeves, D., Gandara, J., Chhangawala, S., & Ahsanuddin, S. (2015). Geospatial resolution of human and bacterial diversity with cityscale metagenomics. *Cell Systems, 1*(1), 72–87.

Alonso, A., Marsal, S., & Julià, A. (2015). Analytical methods in untargeted metabolomics: State of the art in 2015. *Frontiers in Bioengineering and Biotechnology, 3*, 23.

Amberg, A., Riefke, B., Schlotterbeck, G., Ross, A., Senn, H., Dieterle, F., & Keck, M. (2017). *NMR and MS methods for metabolomics*. New York, NY: Humana Press.

Ambroise, C., McLachlan Geoffrey, J., & Do Kim-Anh, Christopher (2005). *Analyzing microarray gene expression data*. Hoboken: John Wiley & Sons, ISBN 9780471726128.

Anderson, J. R., Jones, B. W., Watt, C. B., Shaw, M. V., Yang, J. H., Demill, D., Lauritzen, J. S., Lin, Y., Rapp, K. D., Mastronarde, D., Koshevoy, P., et al. (2011). Exploring the retinal connectome. *Molecular Vision, 17*, 355–379.

Archie, A., & Tung, J. (2015). Social behavior and the microbiome. *Current Opinion in Behavioral Sciences, 6*, 28–34.

Arnold, J. W., Roach, J., & Azcarate-Peril, M. A. (2016). Emerging technologies for gut microbiome research. *Trends in Microbiology, 24*(11), 887–901.

Auburn, R. P., Kreil, D. P., Meadows, L. A., Fischer, B., Matilla, S. S., & Russell, S. (2005). Robotic spotting of cDNA and oligonucleotide microarrays. *Trends in Biotechnology, 23*, 374–379. https://doi.org/10.1016/j.tibtech.2005.04.002

Bäckhed, F., Ley, R. E., Sonnenburg, J. L., Peterson, D. A., & Gordon, J. I. (2005). Host-bacterial mutualism in the human intestine. *Science, 307*, 1915–1920. https://doi.org/10.1126/science.1104816. PMID: 15790844.

Bainbridge, M. N., Warren, R. L., Hirst, M., Romanuik, T., Zeng, T., Go, A., Delaney, A., Griffith, M., Hickenbotham, M., Magrini, V., & Mardis, E. R. (2006). Analysis of the prostate cancer cell line LNCaP transcriptome using a sequencing-by-synthesis approach. *BMC Genomics, 7*, 246. https://doi.org/10.1186/1471-2164-7-246

Barbulovic-Nad, I., Lucente, M., Sun, Y., Zhang, M., Wheeler, A. R., & Bussmann, M. (2006). Bio-microarray fabrication techniques—A review. *Critical Reviews in Biotechnology, 26*, 237–259. https://doi.org/10.1080/07388550600978358

Barrios-Rodiles, M., Brown, K. R., Ozdamar, B., Bose, R., Liu, Z., Donovan, R. S., Shinjo, F., Liu, Y., Dembowy, J., Taylor, I. W., & Luga, V. (2005). High-throughput mapping of a dynamic signaling network in mammalian cells. *Science, 307*(5715), 1621–1625.

Baxter, I., Hosmani, P. S., Rus, A., Lahner, B., Borevitz, J. O., Muthukumar, B., Mickelbart, M. V., Schreiber, L., Franke, R. B., & Salt, D. E. (2009). Root suberin forms an extracellular barrier that affects water relations and mineral nutrition in Arabidopsis. *PLoS Genetics, 5*(5), e1000492. https://doi.org/10.1371/journal.pgen.1000492

Beger, R. D., Dunn, W., Schmidt, M. A., Gross, S. S., Kirwan, J. A., Cascante, M., Brennan, L., Wishart, D. S., Oresic, M., Hankemeier, T., & Broadhurst, D. I. (2016). Metabolomics enables precision medicine: 'A white paper, community perspective. *Metabolomics, 12*, 149.

Beger, R., Holland, R., Sun, J., Schnackenberg, L., Moore, P., Dent, C., Devarajan, P., & Portilla, D. (2008). Metabonomics of acute kidney injury in children after cardiac surgery. *Pediatric Nephrology, 23*, 977–984. https://doi.org/10.1007/s00467-008-0756-7

Beja, O., Aravind, L., Koonin, E. V., Suzuki, M. T., Hadd, A., Nguyen, L. P., Jovanovich, S. B., Gates, C. M., Feldman, R. A., Spudich, J. L., Spudich, E. N., & DeLong, E. F. (2000). Bacterial rhodopsin: Evidence for a new type of phototrophy in the sea. *Science, 289*(5486), 1902–1906. https://doi.org/10.1126/science.289.5486.1902

Bhalla, R., Narasimhan, K., & Swarup, S. (2005). Metabolomics and its role in understanding cellular responses in plants. *Plant Cell Reports, 24*(10), 562–571.

Bino, R. J., Hall, R. D., Fiehn, O., Kopka, J., Saito, K., Draper, J., Nikolau, B. J., Mendes, P., Roessner-Tunali, U., Beale, M. H., & Trethewey, R. N. (2004). Potential of metabolomics as a functional genomics tool. *Trends in Plant Science, 9*(9), 418–425.

Bota, M., Dong, H. W., & Swanson, L. W. (2012). Combining collation and annotation efforts toward completion of the rat and mouse connectomes in BAMS. *Frontiers in Neuroinformatics, 6,* 2.

Breitbart, M., Hewson, I., Felts, B., Mahaffy, J. M., Nulton, J., Salamon, P., Rohwer, F., et al. (2003). Metagenomic analyses of an uncultured viral community from human feces. *Journal of Bacteriology, 185*(20), 6220–6223.

Bullmore, E., & Sporns, O. (2009). Complex brain networks: Graph theoretical analysis of structural and functional systems. *Nature Reviews Neuroscience, 10,* 186–198.

Burckstummer, T., Bennett, K. L., Preradovic, A., Schütze, G., Hantschel, O., Superti-Furga, G., & Bauch, A. (2006). An efficient tandem affinity purification procedure for interaction proteomics in mammalian cells. *Nature Methods, 3*(12), 1013–1019.

Carraro, S., Giordano, G., Reniero, F., Perilongo, G., & Baraldi, E. (2009). Metabolomics a new frontier for research in pediatrics. *The Journal of Pediatrics, 154,* 638–644. https://doi.org/10.1016/j.jpeds.2009.01.014

Chao, D. Y., Gable, K., Chen, M., Baxter, I., Dietrich, C. R., Cahoon, E. B., Guerinot, M. L., Lahner, B., Lu, S., Markham, J. E., Morrissey, J., Han, G., Gupta, S. D., Harmon, J. M., Jaworski, J. G., Dunn, T. M., & Salt, D. E. (2011). Sphingolipids in the root play an important role in regulating the leaf ionome in *Arabidopsis thaliana. Plant Cell, 23,* 1061–1081.

Chatr-aryamontri, A., Ceol, A., Palazzi, L. M., Nardelli, G., Schneider, M. V., Castagnoli, L., & Cesareni, G. (2007). Mint: The molecular interaction database. *Nucleic Acids Research, 35*(Database issue), D572–D574.

Chen, Z., Shinano, T., Ezawa, T., Wasaki, J., Kimura, K., Osaki, M., & Zhu, Y. (2009a). Elemental interconnections in *Lotus japonicus*: A systematic study of the affects of elements additions on different natural variants. *Soil Science and Plant Nutrition, 55,* 91–101. https://doi.org/10.1111/j.1747 0765.2008.00311.x

Chen, Z., Watanabe, T., Shinano, T., Okazaki, K., & Mitsuru, O. (2009b). Rapid characterization of plant mutants with an altered ion-profile: A case study using *Lotus japonicus. New Phytologist, 181,* 795–801.

Chetwynd, A. J., Dunn, W. B., & Rodriguez-Blanco, G. (2017). Collection and preparation of clinical samples for metabolomics. *Advances in Experimental Medicine and Biology, 965,* 19–44.

Chiang, A. S., Lin, C. Y., Chuang, C. C., Chang, H. M., Hsieh, C. H., Yeh, C. W., Shih, C. T., Wu, J. J., Wang, G. T., Chen, Y. C., Wu, C. C.,

Chen, G. Y., Ching, Y. T., Lee, P. C., Lin, H. H., Hsu, H. W., Huang, Y. A., Chen, J. Y., Chiang, H. J., ... Hwang, J. K. (2011). Three-dimensional reconstruction of brain-wide wiring networks in Drosophila at single-cell resolution. *Current Biology, 21*, 1–11.

Chu, Y., & Corey, D. R. (2012). RNA sequencing: Platform selection, experimental design, and data interpretation. *Nucleic Acid Therapeutics, 22*(4), 271–274. https://doi.org/10.1089/nat.2012.0367. PMC 3426205.

Cook, S. Fowler, Fiehn, O., & Thomashow, M. F. (2004). A prominent role for the CBF cold response pathway in configuring the low-temperature metabolome of Arabidopsis. *Proceedings of the National Academy of Sciences of the United States of America, 101*(42), 15243–15248.

Dan Corlan, Alexandru (2004). *Medline trend: Automated yearly statistics of PubMed results for any query* [Internet]. [cited 2017 Apr 27] http://dan.corlan.net/medline-trend.html.

Daniel, R. (2004). The soil metagenome—A rich resource for the discovery of novel natural products. *Current Opinion in Biotechnology, 15*, 199–204.

Danku, J. M., Lahner, B., Yakubova, E., & Salt, D. E. (2013). Large-Scale Plant Ionomics. In F. J. M. Maathuis (Ed.), *Plant Mineral Nutrients* (pp. 255–276). York, UK: University of York. https://doi.org/10.1007/978-1-62703-152-3_17.

Desikan, R. S., Segonne, F., Fischl, B., Quinn, B. T., Dickerson, B. C., Blacker, D., Buckner, R. L., Dale, A. M., Maguire, R. P., Hyman, B. T., Albert, M. S., & Killiany, R. J. (2006). An automated labeling system for subdividing the human cerebral cortex on MRI scans into gyral based regions of interest. *NeuroImage, 31*, 968–980.

Dittmer, T. A., Sahni, N., Kubben, N., Hill, D. E., Vidal, M., Burgess, R. C., Roukos, V., & Misteli, T. (2014). Systematic identification of pathological lamin A interactors. *Molecular Biology of the Cell, 25*(9), 1493–1510.

Dosenbach, N. U., Nardos, B., Cohen, A. L., Fair, D. A., Power, J. D., Church, J. A., Nelson, S. M., Wig, G. S., Vogel, A. C., LessovSchlaggar, C. N., Barnes, K. A., Dubis, J. W., Feczko, E., Coalson, R. S., Pruett, J. R., Jr., Barch, D. M., Petersen, S. E., & Schlaggar, B. L. (2010). Prediction of individual brain maturity using fMRI. *Science, 329*, 1358–1361.

Eickhoff, S. B., Stephan, K. E., Mohlberg, H., Grefkes, C., Fink, G. R., Amunts, K., & Zilles, K. (2005). A new SPM toolbox for combining probabilistic cytoarchitectonic maps and functional imaging data. *NeuroImage, 25*, 1325–1335.

Eloe-Fadrosh, A., Ivanova, N. N., Woyke, T., & Kyrpides, N. C. (2016). Metagenomics uncovers gaps in amplicon-based detection of microbial diversity. *Nature Microbiology, 1*(4).

Emwas, A. H., Al-Talla, Z. A., Yang, Y., & Kharbatia, N. M. (2015a). Gas chromatography-mass spectrometry of biofluids and extracts. *Methods in Molecular Biology, 1277*, 91.

Emwas, A. H., Luchinat, C., Turano, P., Tenori, L., Roy, R., Salek, R. M., et al. (2015b). Standardizing the experimental conditions for using urine in NMR-based metabolomic studies with a particular focus on diagnostic studies: A review. *Metabolomics, 11*(4), 872–894.

Eyckerman, S., Verhee, A., der Heyden, J. V., Lemmens, I., Ostade, X. V., Vandekerckhove, J., & Tavernier, J. (2001). Design and application of a cytokine-receptor based interaction trap. *Nature Cell Biology, 3*(12), 1114–1119.

Fang, G., Bhardwaj, N., Robilotto, R., & Gerstein, M. B. (2010). Getting started in gene orthology and functional analysis. *PLoS Computational Biology, 6*(3), e1000703.

Fauland, A., Köfeler, H., Trötzmüller, M., Knopf, A., Hartler, J., Eberl, A., Chitraju, C., Lankmayr, E., & Spener, F. (2011). A comprehensive method for lipid profiling by liquid chromatography-ion cyclotron resonance mass spectrometry. *Journal of Lipid Research, 52*, 2314–2322.

Fiehn, O. (2002). Metabolomics—The link between genotypes and phenotypes. *Plant Molecular Biology, 48*(1–2), 155–171.

Fiehn, O., & Weckwerth, W. (2003). Deciphering metabolic networks. *European Journal of Biochemistry, 270*(4), 579–588.

Fields, S., & Song, O. (1989). A novel genetic system to detect protein-protein interactions. *Nature, 340*(6230), 245–246.

Figeys, D. (2008). Mapping the human protein interactome. *Cell Research, 18*(7), 716–724.

Fornito, A., & Bullmore, E. T. (2010). What can spontaneous fluctuations of the blood oxygenation level-dependent signal tell us about psychiatric disorders? *Current Opinion in Psychiatry, 23*, 239–249.

Fornito, A., & Bullmore, E. T. (2014). *Connectomics: A new paradigm for understanding brain disease.* European Neuropsychopharmacology. https://doi.org/10.1016/j.euroneuro.2014.02.011

Fornito, A., Zalesky, A., & Breakspear, M. (2013). Graph analysis of the human connectome: Promise, progress, and pitfalls. *NeuroImage, 80C*, 426–444.

Gamazon, E. R., Wheeler, H. E., Shah, K. P., Mozaffari, S. V., Aquino-Michaels, K., Carroll, R. J., Eyler, A. E., Denny, J. C., Nicolae, D. L., Cox, N. J., & Im, H. K. (2015). A gene-based association method for mapping traits using reference transcriptome data. *Nature Genetics, 47*, 1091–1098.

Gamez-Pozo, A., Sanchez-Navarro, I., Calvo, E., Agulló-Ortuño, M. T., López-Vacas, R., Díaz, E., Camafeita, E., Nistal, M., Madero, R., Espinosa, E., & López, J. A. (2012). PTRF/cavin-1 and MIF proteins are identified as non-small cell lung cancer biomarkers by label-free proteomics. *PLoS One, 7*(3), e33752.

Gika, H. G., Theodoridis, G. A., Plumb, R. S., & Wilson, I. D. (2014). Current practice of liquid chromatography-mass spectrometry in metabolomics and metabonomics. *Journal of Pharmaceutical and Biomedical Analysis, 87*, 12–25.

Gilbert, J. A., Quinn, R. A., Debelius, J., Xu, Z. Z., Morton, J., Garg, N., Jansson, J. K., Dorrestein, P. C., & Knight, R. (2016). Microbiome wide association studies link dynamic microbial consortia to disease. *Nature, 535*(7610), 94–103.

Gillespie, E., Brady, S. F., Bettermann, A. D., Cianciotto, N. P., Liles, M. R., Rondon, M. R., Clardy, J., Goodman, R. M., & Handelsman, J. (2002). Isolation of antibiotics turbomycin a and B from a metagenomic library of soil microbial DNA. *Applied and Environmental Microbiology, 68*(9), 4301–4306.

Gill, S. R. M., Pop, R. T., DeBoy, Eckburg, P. B., Turnbaugh, P. J., Samuel, B. S., Gordon, J. I., Relman, D. A., Fraser-Liggett, C. M., & Nelson, K. E. (2006). Metagenomic analysis of the human distal gut microbiome. *Science, 312*(5778), 1355–1359.

Glasser, M. F., & Van Essen, D. C. (2011). Mapping human cortical areas in vivo based on myelin content as revealed by T1- and T2-weighted MRI. *Journal of Neuroscience, 31*, 11597–11616.

Goodwin, S., McPherson, J. D., & McCombie, W. R. (2016). Coming of age: Ten years of next-generation sequencing technologies. *Nature Reviews Genetics, 17*, 333–351. https://doi.org/10.1038/nrg.2016.49. PMID: 27184599.

Gorelick, R., & Heng, H. H. Q. (2011). Sex reduces genetic variation: A multidisciplinary review. *Evolution, 65*(4), 1088–1098.

Gowda, G. A., & Djukovic, D. (2014). Overview of mass spectrometry-based metabolomics: Opportunities and challenges. *Methods in Molecular Biology, 1198*, 3–12.

Griffa, A., Baumann, P. S., Thiran, J.-P., & Hagmann, P. (2013). Structural connectomics in brain diseases. *Neuroimage, 80*, 515–526. https://doi.org/10.1016/j.neuroimage.2013.04.056

Griffin, J. L., Atherton, H., Shockcor, J., & Atzori, L. (2011). Metabolomics as a tool for cardiac research. *Nature Reviews Cardiology, 8*(11), 630–643.

Griffin, J. L., & Shockcor, J. P. (2004). Metabolic profiles of cancer cells. *Nature Reviews Cancer, 4*(7), 551–561.

Guazzaroni, M. E., Beloqui, A., Golyshin, P. N., & Ferrer, M. (2009). Metagenomics as a new technological tool to gain scientific knowledge. *World Journal of Microbiology and Biotechnology, 25*(6), 945–954.

Guazzaroni, M. E., Silva-Rocha, R., & Ward, R. J. (2015). Synthetic biology approaches to improve biocatalyst identification in metagenomic library screening. *Microbial Biotechnology, 8*(1), 52–64.

Gu, R., Chen, F., Liu, B., Wang, X., Liu, J., Li, P., Pan, Q., Pace, J., Soomro, A. A., Lübberstedt, T., & Mi, G. (2015). Comprehensive phenotypic analysis and quantitative trait locus identification for grain mineral concentration, content, and yield in maize (Zea mays L.). *Theoretical and Applied Genetics, 128*, 1777–1789.

Gusev, A., Mancuso, N., Won, H., Kousi, M., Finucane, H. K., Reshef, Y., Song, L., Safi, A., McCarroll, S., Neale, B. M., & Ophoff, R. A. (2018). Transcriptome-wide association study of schizophrenia and chromatin activity yields mechanistic disease insights. *Nature Genetics, 50*, 538–548.

Haga, A., Ogawara, Y., Kubota, D., Kitabayashi, I., Murakami, Y., & Kondo, T. (2013). Interactomic approach for evaluating nucleophosmin binding proteins as biomarkers for Ewing's sarcoma. *Electrophoresis, 34*(11), 1670–1678.

Hagmann, P., Kurant, M., Gigandet, X., Thiran, P., Wedeen, V. J., Meuli, R., & Thiran, J. P. (2007). Mapping human whole-brain structural networks with diffusion MRI. *PLoS One, 2*, e597.

Hanahan, D., & Weinberg, R. A. (2000). The hallmarks of cancer. *Cell, 100*(1), 57–70.

Handelsman, J. (2005). Metagenomics: Application of genomics to uncultured microorganisms. *Microbiology and Molecular Biology Reviews, 69*(1), 195.

Handelsman, J., Rondon, M. R., Brady, S. F., Clardy, J., & Goodman, R. M. (1998). Molecular biological access to the chemistry of unknown soil microbes: A new frontier for natural products. *Chemistry & Biology, 5*(10), R245–R249.

Han, X., Yang, K., & Gross, R. W. (2012). Multi-dimensional mass spectrometry-based shotgun lipidomics and novel strategies for lipidomic analyses. *Mass Spectrometry Reviews, 31*, 134–178.

Hashimshony, T., Wagner, F., Sher, N., & Yanai, I. (2012). CEL-seq: Single-cell RNA-seq by multiplexed linear amplification. *Cell Reports, 2*, 666–673. https://doi.org/10.1016/j.celrep.2012.08.00

Heller, M. J. (2002). DNA microarray technology: Devices, systems, and applications. *Annual Review of Biomedical Engineering, 4*, 129–153.

Hu, Y., Li, M., Lu, Q., Weng, H., Wang, J., Zekavat, S. M., Yu, Z., Li, B., Gu, J., Muchnik, S., & Shi, Y. (2019). A statistical framework for cross-tissue transcriptome-wide association analysis. *Nature Genetics, 51*, 568–576.

Irizarry, R. A., Bolstad, B. M., Collin, F., Cope, L. M., Hobbs, B., & Speed, T. P. (2003). Summaries of affymetrix gene chip probe level data. *Nucleic Acids Research, 31*, e15.

Islam, S., Zeisel, A., Joost, S., La Manno, G., Zajac, P., Kasper, M., Lönnerberg, P., & Linnarsson, S. (2014). Quantitative single-cell RNA-seq with unique molecular identifiers. *Nature Methods, 11*, 163–166. https://doi.org/10.1038/nmeth.2772 PMID:24363023

Iturria-Medina, Y., Canales-Rodriguez, E. J., Melie-Garcia, L., Valdes-Hernandez, P. A., Martinez Montes, E., Aleman-Gomez, Y., & Sanchez-Bornot, J. M. (2007). Characterizing brain anatomical connections using diffusion weighted MRI and graph theory. *NeuroImage, 36*, 645–660.

Ivanov, A. S., Zgoda, V. G., & Archakov, A. I. (2011). Technologies of protein interactomics: A review. *Russian Journal of Bioorganic Chemistry, 37*(1), 4–16.

Kau, L., Ahern, P. P., Griffin, N. W., Goodman, A. L., & Gordon, J. I. (2011). Human nutrition, the gut microbiome and the immune system. *Nature, 474*(7351), 327–336.

Kelley, J. M., Gocayne, J. D., Dubnick, M., Polymeropoulos, M. H., Xiao, H., Merril, C. R., Merril, C. R, Olde, B., Moren, R. F., & Kerlavage, A. R. (1991). Complementary DNA sequencing: Expressed sequence tags and human genome project. *Science, 252*, 1651–1656. pmid:2047873.

Keshava Prasad, T. S., Goel, R., Kandasamy, K., Keerthikumar, S., Kumar, S., Mathivanan, S., Telikicherla, D., Raju, R., Shafreen, B., Venugopal, A., & Balakrishnan, L. (2009). Human protein reference database— 2009 update. *Nucleic Acids Research, 37*(Database issue), D767–D772.

Kim, K. M., Darst, B. F., Deming, Y. K., Zhong, X., Wu, Y., Kang, H., Carlsson, C. M., Johnson, S. C., Asthana, S., & Engelman, C. D. (2021). Cerebrospinal fluid metabolomics identifies 19 brain related phenotype associations. *Communications Biology, 4*, 63. https://doi.org/10.1038/s42003-020 01583-z

Kivioja, T., Vähärautio, A., Karlsson, K., Bonke, M., Enge, M., Linnarsson, S., & Taipale, J. (2011). Counting absolute numbers of molecules using unique molecular identifiers. *Nature Methods, 9,* 72–74. https://doi.org/10.1038/nmeth.1778 PMID: 22101854

Klindworth, E., Pruesse, T., Schweer, Peplies, J., Quast, C., Horn, M., & Glöckner, F. O. (2013). Evaluation of general 16S ribosomal RNA gene PCR primers for classical and next-generation sequencing-based diversity studies. *Nucleic Acids Research, 41*(1), e1–e11.

Knierim, E., Lucke, B., Schwarz, J. M., Schuelke, M., & Seelow, D. (2011). Systematic comparison of three methods for fragmentation of long-range PCR products for next generation sequencing. *PLoS One, 6,* e28240. https://doi.org/10.1371/journal.pone.0028240 PMID: 22140562

Kolch, W., & Pitt, A. (2010). Functional proteomics to dissect tyrosine kinase signalling pathways in cancer. *Nature Reviews Cancer, 10*(9), 618–629.

Korwar, A. M., Bhonsle, H. S., Ghole, V. S., Kachru, R., Gawai, Chaitanyananda, B., Koppikar, & Mahesh, J., Kulkarni (2013). Proteomic profiling and interactome analysis of ER positive/HER2/neu negative invasive ductal carcinoma of the breast: Towards proteomics biomarkers. *Omics, 17*(1), 27–40.

Lahner, B., Gong, J., Mahmoudian, M., Smith, E. L., Abid, K. B., Rogers, E. E., Guerinot, M. L., Harper, J. F., Ward, J. M., McIntyre, L., Schroeder, J. I., & Salt, D. E. (2003). Genomic scale profiling of nutrient and trace elements in *Arabidopsis thaliana. Nature Biotechnology, 21,* 1215–1221.

Lee, J. H., Daugharthy, E. R., Scheiman, J., Kalhor, R., Yang, J. L., Ferrante, T. C., Terry, R., Jeanty, S. S., Li, C., Amamot, R., & Peters, D. T. (2014). Highly multiplexed subcellular RNA sequencing in situ. *Science, 343,* 1360–1363. https://doi.org/10.1126/science.1250212

Levin, J. Z., Yassour, M., Adiconis, X., Nusbaum, C., Thompson, D. A., Friedman, N., Gnirke, A., & Regev, A. (2010). Comprehensive comparative analysis of strand-specific RNA sequencing methods. *Nature Methods, 7,* 709–715. https://doi.org/10.1038/nmeth.1491. PMID: 20711195.

Lievens, S., Eyckerman, S., Lemmens, I., & Tavernier, J. (2010). Large-scale protein interactome mapping: Strategies and opportunities. *Expert Review of Proteomics, 7*(5), 679–690.

Lievens, S., Gerlo, S., Lemmens, I., De Clercq, D. J., Risseeuw, M. D., Vanderroost, N., De Smet, A. S., Ruyssinck, E., Chevet, E., Van Calenbergh, S., & Tavernier, J. (2014). KISS, a mammalian in situ protein interaction sensor. *Molecular & Cellular Proteomics, 13*(12), 3332–3342.

Lin, T. W., Wu, J. W., & Chang, D. T. (2013). Combining phylogenetic profiling-based and machine learning based techniques to predict functional related proteins. *PLoS One, 8*(9), e75940.

Liu, T., Xue, R., Dong, L., Wu, H., Zhang, D., & Shen, X. (2011). Rapid determination of serological cytokine biomarkers for hepatitis B virus-related hepatocellular carcinoma using antibody microarrays. *Acta Biochimica et Biophysica Sinica, 43*(1), 45–51.

Lloyd-Price, J., Mahurkar, A., Rahnavard, G., Crabtree, J., Orvis, J., Hall, A. B., Brady, A., Creasy, H. H., McCracken, C., Giglio, M. G., & McDonald, D. (2017). Strains, functions and dynamics in the expanded human microbiome project. *Nature, 550*, 61–66.

Lockhart, D. J., Dong, H., Byrne, M. C., Follettie, M. T., Gallo, M. V., Chee, M. S., Mittmann, M., & Wang, C. (1996). Expression monitoring by hybridization to high-density oligonucleotide arrays. *Nature Biotechnology, 14*, 1675–1680. https://doi.org/10.1038/nbt1296-1675

Loman, N. J., Misra, R. V., Dallman, T. J., Constantinidou, C., Gharbia, S. E., Wain, J., & Pallen, M. J. (2012). Performance comparison of benchtop high-throughput sequencing platforms. *Nature Biotechnology, 30*, 434–439. https://doi.org/10.1038/nbt.2198 PMID: 22522955

Louca, S., Polz, M. F., Mazel, F., Albright, M. B., Huber, J. A., O'Connor, M. I., Ackermann, M., Hahn, A. S., Srivastava, D. S., Crowe, S. A., & Doebeli, M. (2018). Function and functional redundancy in microbial systems. *Nature Ecology & Evolution, 2*(6), 936–943.

Maher, C. A., Kumar-Sinha, C., Cao, X., Kalyana-Sundaram, S., Han, B., Jing, X., et al. (2009). Transcriptome sequencing to detect gene fusions in cancer. Bibcode:2009Natur.458...97M *Nature, 458*(7234), 97–101. https://doi.org/10.1038/nature07638. PMC 2725402. PMID 19136943.

Maier, T. V., & Schmitt-Kopplin, P. (2016). Capillary electrophoresis in metabolomics. *Methods in Molecular Biology, 1483*, 437–470.

Markley, J. L., Brüschweiler, R., Edison, A. S., Eghbalnia, H. R., Powers, R., Raftery, D., & Wishart, D. S. (February 2017). The future of NMR-based metabolomics. *Current Opinion in Biotechnology, 43*, 34–40.

Maskrey, B. H., et al. (2013). Emerging importance of omega-3 fatty acids in the innate immune response: Molecular mechanisms and lipidomic strategies for their analysis. *Molecular Nutrition & Food Research, 57*, 1390–1400.

Masuo, Y., Imai, T., Shibato, J., et al. (2009). Omic analyses unravels global molecular changes in the brain and liver of a rat model for chronic Sake

(Japanese alcoholic beverage) intake. *Electrophoresis, 30*(8), 1259−1275.

Melé, M., Ferreira, P. G., Reverter, F., DeLuca, D. S., Monlong, J., Sammeth, M., et al. (2015). Human genomics. The human transcriptome across tissues and individuals. *Science, 348*, 660−665. https://doi.org/10.1126/science.aaa0355

Mendes, P., Kell, D. B., & Westerhoff, H. V. (1992). Channelling can decrease pool size. *European Journal of Biochemistry, 204*(1), 257−266.

Menedes, P., Kell, D. B., & Westerhoff, H. V. (1996). Why and when channeling can decrease pool size at constant net flux in a simple dynamic channel. *Biochimica et Biophysica Acta, 1289*(2), 175−186.

Meyers, B. C., Vu, T. H., Tej, S. S., Ghazal, H., Matvienko, M., Agrawal, V., et al. (2004). Analysis of the transcriptional complexity of *Arabidopsis thaliana* by massively parallel signature sequencing. *Nature Biotechnology, 22*, 1006−1011. https://doi.org/10.1038/nbt992

Modha, D. S., & Singh, R. (2010). Network architecture of the long distance pathways in the macaque brain. *Proceedings of the National Academy of Sciences of the United States of America, 107*, 13485−13490.

Morozova, O., Hirst, M., & Marra, M. A. (2009). Applications of new sequencing technologies for transcriptome analysis. *Annual Review of Genomics and Human Genetics, 10*, 135−151. https://doi.org/10.1146/annurev genom-082908-145957

Moult, J., Fidelis, K., Kryshtafovych, A., & Tramontano, A. (2011). Critical assessment of methods of protein structure prediction (CASP)−round IX. *Proteins, 79*(Suppl. 10), 1−5.

Nelson, N. J. (April 4, 2001). Microarrays have arrived: Gene expression tool matures. *J Natl Cancer Inst, 93*(7), 492−494.

Nicol, G. W., & Schleper, C. (2006). Ammonia-oxidising crenarchaeota: Important players in the nitrogen cycle? *Trends in Microbiology, 14*(5), 207−212. https://doi.org/10.1016/j.tim.2006.03.004

Norton, G. J., Deacon, C. M., Xiong, L. Z., Huang, S. Y., Meharg, A. A., & Price, A. H. (2010). Genetic mapping of the rice ionome in leaves and grain: Identification of QTLs for 17 elements including arsenic, cadmium, iron and selenium. *Plant and Soil, 329*, 139−153.

Ogmen, U., Keskin, O., Aytuna, A. S., et al. (2005). Prism: Protein interactions by structural matching. *Nucleic Acids Research, 33*(Web Server issue), W331−W336.

Oliver, S. G., Winson, M. K., Kell, D. B., & Baganz, F. (1998). Systematic functional analysis of the yeast genome. *Trends in Biotechnology, 16*, 373−378.

Osburne, M. S., Grossman, T. H., August, P. R., & MacNeil, I. A. (2000). Tapping into microbial diversity for natural products drug discovery. *ASM News, 66*, 411–417.

Oulas, C., Pavloudi, P., Polymenakou, et al. (2015). Metagenomics: Tools and insights for analyzing next-generation sequencing data derived from biodiversity studies. *Bioinformatics and Biology Insights, 9*. BBI.S12462–BBI.S12488.

Ozsolak, F., & Milos, P. M. (2011). RNA sequencing: Advances, challenges and opportunities. *Nature Reviews Genetics, 12*, 87–98. https://doi.org/10.1038/nrg2934

Pace, N. R., Stahl, D. A., Lane, D. J., & Olsen, G. J. (1986). The analysis of natural microbial populations by ribosomal RNA sequences. In M. K. Cou (Ed.), *55, Advances in microbial ecology* (p. 1). Boston, MA, USA: Springer.

Paine, J. A., Shipton, C. A., Chaggar, S., Howells, R. M., Kennedy, M. J., Vernon, G., Wright, S. Y., Hinchliffe, E., Adams, J. L., Silverstone, A. L., & Drake, R. (2005). Improving the nutritional value of Golden Rice through increased pro-vitamin A content. *Nature Biotechnology, 23*(4), 482–487.

Pai, J. K., Pischon, T., Ma, J., Manson, J. E., Hankinson, S. E., Joshipura, K., Curhan, G. C., Rifai, N., Cannuscio, C. C., Stampfer, M. J., & Rimm, E. B. (2004). Inflammatory markers and the risk of coronary heart disease in men and women. *New England Journal of Medicine, 351*, 2599–2610.

Pagel, P., Kovac, S., Oesterheld, M., Brauner, B, Dunger-Kaltenbach, I., Frishman, G., Montrone, C, Mark, P., Stümpflen, V, Mewes, H. W., & Ruepp, A. (2005). The MIPS mammalian protein-protein interaction database. *Bioinformatics, 21*(6), 832–834.

Pan, Q., Shai, O., Lee, L. J., Frey, B. J., & Blencowe, B. J. (2008). Deep surveying of alternative splicing complexity in the human transcriptome by high-throughput sequencing. *Nature Genetics, 40*, 1413–1415. https://doi.org/10.1038/ng.259

Parekh, S., Ziegenhain, C., Vieth, B., Enard, W., & Hellmann, I. (2016). The impact of amplification on differential expression analyses by RNA-seq. *Scientific Reports, 6*, 25533. https://doi.org/10.1038/srep25533. PMID: 27156886.

Petschnigg, J., Snider, J., & Stagljar, I. (2011). Interactive proteomics research technologies: Recent applications and advances. *Current Opinion in Biotechnology, 22*(1), 50–58.

Pieragostino, D., D'Alessandro, M., di Ioia, M., Di Ilio, C., Sacchetta, P., & Del Boccio, P. (2015). Unraveling the molecular repertoire of tears as a source of biomarkers: Beyond ocular diseases. *Proteomics: Clinical Applications, 9*, 169–186.

Porras, P., Aranda, B., Hermjakob, H., & Orchard, S. E. (2012). Analyzing protein–protein interaction networks. *Journal of Proteome Research, 11*(4), 2014–2031. https://doi.org/10.1021/pr201211w

Pozhitkov, A. E., Tautz, D., & Noble, P. A. (2007). Oligonucleotide microarrays: Widely applied—Poorly understood. *Briefings in Functional Genomics and Proteomics, 6*(2), 141–148. https://doi.org/10.1093/bfgp/elm014

Rao, V. S., Srinivas, K., Sujini, G. N., & Kumar, G. N. (2014). Protein-protein interaction detection: Methods and analysis. *International Journal of Proteomics, 2014*, 147648.

Rigaut, G., Shevchenko, A., Rutz, B., Wilm, M., Mann, M., & Séraphin, B. (1999). A generic protein purification method for protein complex characterization and proteome exploration. *Nature Biotechnology, 17*(10), 1030–1032.

Roemer, T., & Boone, C. (2013). Systems-level antimicrobial drug and drug synergy discovery. *Nature Chemical Biology, 9*, 222–231. https://doi.org/10.1038/nchembio.1205

Romanov, V., Davidoff, S. N., Miles, A. R., Grainger, D. W., Gale, B. K., & Brooks, B. D. (2014). A critical comparison of protein microarray fabrication technologies. *Analyst, 139*, 1303–1326. https://doi.org/10.1039/c3an01577g

Saito, K., Hirai, M. Y., & Yonekura-Sakakibara, K. (2008). Decoding genes with coexpression networks and metabolomics—"Majority report by precogs". *Trends in Plant Science, 13*(1), 36–43.

Sánchez-Rodríguez, E., del Mar Rubio-Wilhelmi, M., Cervilla, L. M., Blasco, B., Rios, J. J., Leyva, R., … Ruiz, J. M. (2010). Study of the ionome and uptake fluxes in cherry tomato plants under moderate water stress conditions. *Plant and soil, 335*(1), 339–347.

Sanger, G. M., Air, B. G., Barrell, Brown, N. L., Coulson, A. R., Fiddes, J. C., Hutchison, C. A., Slocombe, P. M., & Smith, M. (1977). Nucleotide sequence of bacteriophage phi X174 DNA. *Nature, 265*(5596), 687–695.

Schena, M., Shalon, D., Davis, R. W., & Brown, P. O. (October 20, 1995). Quantitative monitoring of gene expression patterns with a complementary DNA microarray. *Science, 270*(5235), 467–470.

Schmeisser, Steele, H., & Streit, W. R. (2007). Metagenomics, biotechnology with non-culturable microbes. *Applied Microbiology and Biotechnology, 75*(5), 955–962.

Schwarz, E., & Bahn, S. (2008). Biomarker discovery in psychiatric disorders. *Electrophoresis, 29*, 2884–2890.

Selzer, R. R., Richmond, T. A., Pofahl, N. J., Green, R. D., Eis, P. S., Nair, P., Brothman, A. R., & Stallings, R. L. (2005). Analysis of chromosome breakpoints in neuroblastoma at sub-kilobase resolution using fine-tiling oligonucleotide array CGH. *Genes, Chromosomes & Cancer, 44*, 305–319. https://doi.org/10.1002/gcc.20243

Shah, S. H., & Newgard, C. B. (April 2015). Integrated metabolomics and genomics: Systems approaches to biomarkers and mechanisms of cardiovascular disease. *Circulation: Cardiovascular Genetics, 8*(2), 410–419.

Shanahan, M., Bingman, V. P., Shimizu, T., Wild, M., & Güntürkün, O. (2013). Large-scale network organization in the avian forebrain: A connectivity matrix and theoretical analysis. *Frontiers in Computational Neuroscience, 7*, 89.

Shi, T. L., Li, Y. X., Cai, Y. D., & Chou, K. C. (2005). Computational methods for protein-protein interaction and their application. *Current Protein & Peptide Science, 6*(5), 443–449.

Simrén, J., Ashton, N. J., Blennow, K., & Zetterberg, H. (2019). An update on fluid biomarkers for neurodegenerative diseases: Recent success and challenges ahead. *Current Opinion in Neurobiology, 61*, 29 39.

Smilowitz, J. T., Air, G. M., Barrell, B. G., Brown, N. L., Coulson, A. R., Fiddes, J. C., Hutchison, C. A., Slocombe, P. M., & Smith, M. (2013). Nutritional lipidomics: Molecular metabolism, analytics, and diagnostics. *Molecular Nutrition & Food Research, 57*, 1319–1335.

Sporns, O., Tononi, G., & Kotter, R. (2005). The human connectome: A structural description of the human brain. *PLoS Computational Biology, 1*, e42.

Spratlin, J. L., Serkova, N. J., & Eckhardt, S. G. (2009). Clinical applications of metabolomics in oncology: A review. *Clinical Cancer Research, 15*(2), 431–440.

Stephan, K. E., Kamper, L., Bozkurt, A., Burns, G. A., Young, M. P., & Kotter, R. (2001). Advanced database methodology for the collation of connectivity data on the macaque brain (CoCoMac). *Philosophical Transactions of the Royal Society, London B: Biological Sciences, 356*, 1159–1186.

Stumpf, M. P., Thorne, T., de Silva, E., Stewart, E., An, H. J. R., Lappe, M., & Wiuf, C. (2008). Estimating the size of the human interactome. *Proceedings of the National Academy of Sciences of the United States of America, 105*(19), 6959–6964.

Stynen, B., Tournu, H., Tavernier, J., & Van Dijck, P. (2012). *Diversity in genetic in vivo methods for protein.*

Su, Z., Fang, H., Hong, H., Shi, L., Zhang, W., Zhang, W., Zhang, Y., Dong, Z., Lancashire, L. J., Bessarabova, M., & Yang, X. (2014). An investigation of biomarkers derived from legacy microarray data for their utility in the RNA-seq era. *Genome Biology, 15*, 523. https://doi.org/10.1186/s13059-014-0523-y

Sultan, M., Schulz, M. H., Richard, H., Magen, A., Klingenhoff, A., Scherf, M., Seifert, M., Borodina, T., Soldatov, A., Parkhomchuk, D., Schmidt, D., O'Keeffe, S., Haas, S., Vingron, M., Lehrach, H., & Yaspo, M. L. (2008). A global view of gene activity and alternative splicing by deep sequencing of the human transcriptome. *Science, 321*, 956–960. https://doi.org/10.1126/science.1160342

Sun, L., Zou, L. X., & Chen, M. J. (March 2017). Make precision medicine work for chronic kidney disease. *Medical Principles and Practice, 26*(2), 101–107.

Sunagawa, S., Coelho, L. P., Chaffron, S., Kultima, J. R., Labadi, K., Salazar, G., Djahanschiri, B., Zeller, G., Mende, D. R., Alberti, A., & Cornejo-Castillo, F. M. (2015). Structure and function of the global ocean microbiome. *Science, 348*(6237). article 1261359.

Tachibana Chris. (2015). Transcriptomics today: Microarrays, RNA-seq, and more. *Science, 349 (6247)*, 544–546.

Tang, Y. (2015). Non-genomic omic techniques. In Y. Tang, M. Sussman, D. Liu, I. Poxton, & J. Schwartzman (Eds.), *Molecular medical microbiology* (pp. 399–406). London: Academic Press.

Tringe, S. G., von Mering, C., Kobayashi, A., Salamov, A. A., Chen, K., Chang, H. W., Podar, M., Short, J. M., Mathur, E. J., Detter, J. C., Bork, P., et al. (2005). Comparative metagenomics of microbial communities. *Science, 308*(5721), 554–557.

Tweeddale, H., Notley-McRobb, L., & Ferenci, T. (1998). Effect of slow growth om metabolism of *Escherichia coli*, as revealed by global metabolite pools ("metabolome") analysis. *Journal of Bacteriology, 180*, 5109–5116.

Tyson, G. W., Chapman, J., Hugenholtz, P., Allen, E. E., Ram, R. J., Richardson, P. M., Solovyev, V. V., Rubin, E. M., Rokhsar, D. S., & Banfield, J. F. (2004). Community structure and metabolism through reconstruction of microbial genomes from the environment. *Nature, 428*(6978), 37 43.

Uchiyama, T., Abe, T., Ikemura, T., & Watanabe, K. (2005). Substrate-induced gene-expression screening of environmental metagenome libraries for isolation of catabolic genes. *Nature Biotechnology, 23*(1), 88−93.

Van den Heuvel, M. P., Stam, C. J., Boersma, M., & Hulshoff Pol, H. E. (2008). Small-world and scale-free organization of voxel-based resting-state functional connectivity in the human brain. *NeuroImage, 43*, 528−539.

Venkatesan, K., Rual, J. F., Vazquez, A., Stelzl, U., Lemmens, I., Hirozane-Kishikawa, T., Hao, T., Zenkner, M., Xin, X., Goh, K. I., & Yildirim, M. A. (2009). An empirical framework for binary interactome mapping. *Nature Methods, 6*(1), 83−90.

Venter, J. C., Remington, K., Heidelberg, J. F., Halpern, A. L., Rusch, D., Eisen, J. A., Wu, D., Paulsen, I., Nelson, K. E., Nelson, W., & Fouts, D. E. (2004). Environmental genome shotgun sequencing of the Sargasso Sea. *Science, 304*(5667), 66−74.

Vieites, J. M., Guazzaroni, M. E., Beloqui, A., Golyshin, P. N., & Ferrer, M. (2009). Metagenomics approaches in systems microbiology. *FEMS Microbiology Reviews, 33*(1), 236−255.

Wainberg, M., Sinnott-Armstrong, N., Mancuso, N., Barbeira, A. N., Knowles, D. A., Golan, D., Ermel, R., Ruusalepp, A., Quertermous, T., Hao, K., & Björkegren, J. L. (2019). Opportunities and challenges for transcriptome-wide association studies. *Nature Genetics, 51*, 592−599.

Wang, Z., Gerstein, M., & Snyder, M. (2009). RNA-seq: A revolutionary tool for transcriptomics. *Nature Reviews Genetics, 10*(1), 57−63. https://doi.org/10.1038/nrg2484

Weckwerth, W. (2003). Metabolomics in systems biology. *Annual Review of Plant Biology, 54*, 669−689.

White, J. G., Southgate, E., Thomson, J. N., & Brenner, S. (1986). The structure of the nervous system of the nematode *Caenorhabditis elegans*. *Philosophical Transactions of the Royal Society, London B: Biological Sciences, 314*, 1−340.

Wishart, D. S., Jewison, T., Guo, A. C., Wilson, M., Knox, C., Liu, Y., Djoumbou, Y., Mandal, R., Aziat, F., Dong, E., & Bouatra, S. (2012). HMDB 3.0—the human metabolome database in 2013. *Nucleic Acids Research, 41*(D1), D801−D807.

Wu, D., Shen, Q., Cai, S., Chen, Z. H., Dai, F., & Zhang, G. (2013). Ionomic responses and correlations between elements and metabolites under salt stress in wild and cultivated barley. *Plant Cell Physiology, 54*, 1976−1988.

Yeo, B. T., Krienen, F. M., Sepulcre, J., Sabuncu, M. R., Lashkari, D., Hollinshead, M., Roffman, J. L., Smoller, J. W., Zollei, L.,

Polimeni, J. R., Fischl, B., Liu, H., & Buckner, R. L. (2011). The organization of the human cerebral cortex estimated by intrinsic functional connectivity. *Journal of Neurophysiology, 106*, 1125−1165.

Young, M. P., Scannell, J. W., Burns, G. A., & Blakemore, C. (1994). Analysis of connectivity: Neural systems in the cerebral cortex. *Reviews in the Neurosciences, 5*, 227−250.

Yu, Y., Tan, P., Zhuang, Z., Wang, Z., Zhu, L., Qiu, R., & Xu, H. (2021). Untargeted metabolomic approach to study the serum metabolites in women with polycystic ovary syndrome. *BMC Medical Genomics, 14*, 206. https://doi.org/10.1186/s12920-021-01058-y

Zafar, S., Asif, A. R., Ramljak, S., Tahir, W., Schmitz, M., & Zerr, I. (2014). Anchorless 23-230 PrPC interactomics for elucidation of PrPC protective role. *Molecular Neurobiology, 49*(3), 1385−1399.

Zalesky, A., & Fornito, A. (2009). A DTI-derived measure of corticocortical connectivity. *IEEE Transactions on Medical Imaging, 28*, 1023−1036.

Zargar, S. M., Nancy, G., Muslima, N., Rakeeb, A. M., Surinder, K. G., Ganesh, K. A., & Randeep, R. (2016). *Omics—A new approach to sustainable production, breeding oilseed crops for sustainable production.* Academic Press. https://doi.org/10.1016/B978-0-12-801309-0.00013-6. ISBN 9780128013090.

Zhang, M., Pinson, S. R., Tarpley, L., Huang, X. Y., Lahner, B., Yakubova, E., Baxter, I., Guerinot, M. L., & Salt, D. E. (2014). Mapping and validation of quantitative trait loci associated with concentrations of 16 elements in unmilled rice grain. *Theoretical and Applied Genetics, 127*, 137−165.

Zhou, L., & Beuerman, R. W. (2012). Tear analysis in ocular surface diseases. *Progress in Retinal and Eye Research, 31*, 527−550.

Zhou, D., & He, Y. (2008). Extracting interactions between proteins from the literature. *Journal of Biomedical Informatics, 41*(2), 393−407.

Ziegler, G., Terauchi, A., Becker, A., Armstrong, P., Hudson, K., & Baxter, I. (2013). Ionomic screening of field-grown soybean identifies mutants with altered seed elemental composition. *The Plant Genome, 6*, 2.

Further reading

Protein interaction studies: from the yeast two-hybrid system to the mammalian split-luciferase system. *Microbiology and Molecular Biology Reviews, 76*(2), (2012), 331−382.

Rodrigues, M. A., Santos, C. A. F., & Santana, J. R. F. (2012). Mapping of AFLP loci linked to tolerance to cowpea golden mosaic virus. *Genetics and Molecular Research, 11*, 3789–3797.

Visioli, F. (2015). Lipidomics to assess omega 3 bioactivity. *Journal of Clinical Medicine, 4*, 1753–1760.

Wedeen, V. J., Wang, R. P., Schmahmann, J. D., Benner, T., Tseng, W. Y. I., Dai, G., Pandya, D. N., Hagmann, P., D'Arceuil, H., & de Crespigny, A. J. (2008). Diffusion spectrum magnetic resonance imaging (DSI) tractography of crossing fibers. *NeuroImage, 41*(4), 1267–1277.

Index

'*Note*: Page numbers followed by "f" indicate figures and "t" indicate tables.'

A

Ab initio and similarity, 61–62
Ab initio gene prediction, 62
Ab initio protein structure predictions, 170–172
Acetylation, 6
Agricultural sciences, 202–203
Amaranth, 100
Amplified fragment length polymorphism
    (AFLP), 36–37
Analytical microarrays, 184–185
Arabidopsis genome, 59
Arbitrarily primed polymerase chain reaction
    (AP-PCR), 34

B

Bacterial cells, 146
Beverages analysis, 146
BioShell, 170
Buckwheat, 99
Buffers, 114

C

Cancer, 212
   cell proteomes, 134
Capillary electrophoresis, 199
Celiac disease, 94–95
Chain-termination sequencing method, 48–49
   basic principle of, 48–49
ChIA-PET, 80
ChIP-CpG microarray, 77–78
ChIP-exo, 79
ChIP-Seq, 78–79
Chromatin, 5–6
Chromatin immunoprecipitation, 73–82, 75f
Chromatin immunoprecipitation (ChIP), 65–66,
    74–76
   ChIA-PET, 80
   ChIP-CpG microarray, 77–78
   ChIP-exo, 79
   ChIP-Seq, 78–79
   chromatin immunoprecipitation-chip-DNA
     microarray technology, 77
   chromosomes, repressed regions of, 82
   cloning, 77
   cross-linking reversal, 76

Encyclopedia of DNA elements (ENCODE)
    project, 81
epigenetic modifications, 82
eukaryotic genomes
   repressed transcription sites in, 82
   transcription sites in, 82
ideal chromatin, 76
lysis of cells, 76
proChIPdb, 80
PRODORIC2 database, 81
protein binding DNA sequence, 76
protein–DNA cross-linking, 74
RedChIP, 79–80
RNA polymerase II, 81–82
target protein–DNA complex, immunoprecipitation of, 76
transcription factors, 81–82
Chromatin immunoprecipitation-chip-DNA
    microarray technology, 77
Chromosomes, repressed regions of, 82
Circular dichroism (CD), 159–161
   applications, 161
   basic workflow, 160
Cloning, 77
Connectomics
   technical inputs for, 222–224
    applications of, 224–225
     magnetic resonance imaging (MRI), 223–224,
     225f
Cross-linking reversal, 76

D

2D gel electrophoresis-combining SDS-PAGE,
    118–120
   basic principle, 118
   fitting and loading first-dimension gel (IEF),
    120
   gel for one-dimension IEF process, 118–119
   preparing and pouring of, 119–120
DNA
   footprinting
    applications, 69
    electrophoretic mobility shift assay/band shift
     assay, 69–73
    gel retardation assay, 69–73
    workflow, 67–69, 68f

DNA (*Continued*)
   RNA analysis, 145
DNA footprinting/deoxyribonuclease I (DNase I),
   66—69, 67f

E
EMSA
   biotinylated DNA, 73
   cancer detection, 73
   DNA, nuclear receptors in, 73
   TF-binding sites, 73
   variants of, 71—72
Encyclopedia of DNA elements (ENCODE)
   project, 81
Environmental factors, 203—204
Epigenetics, 5—7
   chromosomes, 7—9
   DNA methylation, 5
   histone modifications, 5—7
   nuclear and organelle genomes, eukaryotes, 7—10
   nuclear genome, 9—10
   protein level, 5—7
Eukaryotic genome organization, 10—18
Eye diseases, 212—213

F
Fluorescence detection workflow, 147—148
   filter fluorometer, 147—148
   major applications of, 148—149
Fluorescence in situ hybridization (FISH), 41—43
   advantages, 43
   sequence-tagged site mapping, 44
   types of, 41, 42t—43t
Fluorescence spectroscopy, 146—149
   principle, 146—147
Food allergies, 96
Food quality, 102
Functional genomics, 2—3, 3f
Functional protein microarrays, 185—186

G
Gastroesophageal reflux disease (GERD), 95
Gel, 114—115
   interpretation, 116
   running, 116
   visualization of proteins, 116
Generation of models, 167
   fragments assembly, 167
   segments, matching of, 167
Genome annotation, 60—62
Genome mapping
   advantages, 40

amplified fragment length polymorphism
   (AFLP), 36—37
definition, 29—30
disadvantages, 40
inter simple sequence repeats (ISSRs), 32—33
   advantage, 33
   disadvantage, 33, 33f
linkage maps, 29
markers, 30—40, 30f
physical mapping, 40—44
   Fluorescence in situ hybridization, 41—43
   restriction mapping, 40—41
physical maps, 30
randomly amplified polymorphic DNA, 33—35,
   34f
   advantages, 34
   applications, 34
   limitations, 35
restriction fragment length polymorphism, 35—36
   application, 36
simple-sequence repeats (SSRs), 31—32, 32t
single nucleotide polymorphism (SNP), 37—40,
   39f
types of, 30f
Genomes analysis-I
   chain-termination sequencing method, 48—49
      basic principle of, 48—49
   genome sequencing, 47—48
   hierarchical shotgun sequencing, 50—51
   human genome project, 55—57, 56f
      arabidopsis genome, 59
      facts of, 57—59
      genome annotation, 60—62
      microbial genome sequencing, 59—60, 60t
   next-generation sequencing methods, 51—54
      Illumina (Solexa) sequencing, 51
      ion torrent, 52
      Roche 454 sequencing, 52
      sequencing methods, 53—54
      techniques, 51—53
   sangers sequencing, 48—49
   whole-genome shotgun sequencing, 50
Genome sequencing, 47—48
Genomes analysis-II
   basic working principle, 70—71, 71f
   cellular functions, 65—66
   chromatin immunoprecipitation (ChIP), 65—66,
      73—82, 75f
      ChIA-PET, 80
      ChIP-CpG microarray, 77—78
      ChIP-exo, 79
      ChIP-Seq, 78—79

chromatin immunoprecipitation-chip-DNA
microarray technology, 77
chromosomes, repressed regions of, 82
cloning, 77
cross-linking reversal, 76
Encyclopedia of DNA elements (ENCODE)
project, 81
epigenetic modifications, 82
eukaryotic genomes, repressed transcription
sites in, 82
eukaryotic genomes, transcription sites in, 82
ideal chromatin, 76
lysis of cells, 76
proChIPdb, 80
PRODORIC2 database, 81
protein binding DNA sequence, 76
protein—DNA cross-linking, 74
RedChIP, 79—80
RNA polymerase II, 81—82
target protein—DNA complex, immunoprecipitation of, 76
transcription factors, 81—82
DNA footprinting
applications, 69
electrophoretic mobility shift assay/band shift
assay, 69—73
gel retardation assay, 69—73
workflow, 67—69, 68f
DNA footprinting/deoxyribonuclease I (DNase
I), 66—69, 67f
EMSA
biotinylated DNA, 73
cancer detection, 73
DNA, nuclear receptors in, 73
TF-binding sites, 73
variants of, 71—72
Genomics
DNA, 1—2
epigenetics, 5—7
chromosomes, 7—9
DNA methylation, 5
histone modifications, 5—7
nuclear and organelle genomes, eukaryotes,
7—10
nuclear genome, 9—10
protein level, 5—7
eukaryotic genome organization, 10—18
functional genomics, 2—3, 3f
organelle genome editing, 18—20
mitome-mitochondrial DNA, 19—20
plastome-chloroplast DNA, 19
regulatory elements, 12—18
CpG islands, 13

highly repetitive DNA sequences, 15
intergenic/extragenic DNA, 14
introns, 14
long interspersed elements, 18
LTR transposons, 17
moderately repetitive DNA sequences, 16—18
non-LTR transposons, 17—18
nonrepetitive and repetitive DNA sequences,
14—15
other noncoding sequences, 13
others, 13
promoters, 12—13
pseudogenes, 14
5' untranslated region (UTR), 14
RNA, 1—2
structural genomics, 3—4, 4f
structure of, 10—18

H

Haemophilus influenza, 60, 61t
HHpred, 169
Hierarchical shotgun sequencing, 50—51
High-performance liquid chromatography, 199
Histone phosphorylation, 6
Homology/template-based modeling, 166—169
Human genome project, 55—57, 56f
arabidopsis genome, 59
facts of, 57—59
genome annotation, 60—62
microbial genome sequencing, 59—60, 60t
Human nutrition and diseases, 94
Hypertension, 101

I

Illumina (Solexa) sequencing, 51
Inductively coupled plasma (ICP),
220—221
Inductively coupled plasma-atomic emission
spectroscopy (ICP-AES), 220—221
Inductively coupled plasma-optical emission
spectroscopy (ICP-OES), 220—221
In silico interactome analysis, 207—208
Interactomics, 204—209, 204f—205f
applications of, 208—209
biomarker, 209
cellular and disease biology, 208—209
drug targets, 209
in silico interactome analysis, 207—208
in vitro methods, 205—206
in vivo methods, 206—207
Inter simple sequence repeats (ISSRs), 32—33
advantage, 33
disadvantage, 33, 33f

In vitro methods, 205–206
In vivo methods, 206–207
Ionomics
    applications of, 221, 222f
    inductively coupled plasma (ICP), 220–221
    inductively coupled plasma-atomic emission
        spectroscopy (ICP-AES), 220–221
    inductively coupled plasma-optical emission
        spectroscopy (ICP-OES), 220–221
    mass spectrometry, 220–221
    technical intervention for, 216f, 219–221
Ion torrent, 52
Irritable bowel syndrome, 95
Isobaric tags for relative and absolute quantifica-
        tion (iTRAQ), 128–130
    advantages, 130
    applications, 130
    basic principle, 129–130
    disadvantages of, 130
Isoelectric focusing gels (IEF), 117–118
    preparation, 117–118
Isotope-coded affinity tags, 125–127
    applications of, 127
    basic principle, 125–126
    workflow, 125–126
Isotope-coded protein labeling, 127–128
    applications, 128

L
Lifestyle diseases, 96–97, 102
    plant-based foods to, 100–101
Linkage maps, 29
Lipidomics, 209–213
    technical intervention, 210–211
        cancers, 212
        eye diseases, 212–213
        mass spectrometry, 210–211, 211f
        metabolic syndromes, 211–212
        neurological diseases, 212
        nutrition, 213
        screening and discovery, 213

M
Magnetic resonance imaging (MRI), 223–224,
        225f
MALDI-TOF mass spectrometry, 120–125
    bioimaging, 124
    biomarkers, 124
    higher resolution reflectron, 124
    laser types, 122
    liner TOF analyzer, 122
    matrix compounds, 122

microorganisms, 124
polymers, 125
principle and setup of, 120
proteomes, 124
reflectron TOF analyzer, 122
sample concentration for, 121
sampling plates, 121–122
Markers, 30–40, 30f
Mass spectrometry, 210–211, 211f, 220–221
Mass spectroscopy, 199
Metabolic syndromes, 211–212
Metabolomics, 196–204
    agricultural sciences, 202–203
    applications of, 200–204
    environmental factors, 203–204
    in human health and diseases, 201–202
    nutrigenomics, 203
    personalized medicine, 203
    plant biology, 202–203
    technical intervention, 197–199, 198f
        capillary electrophoresis, 199
        collection of samples, 197–199
        high-performance liquid chromatography, 199
        mass spectroscopy, 199
        nuclear magnetic resonance (NMR), 198–199
    workflow for, 199–200
Metagenomics
    applications of, 215–218, 217f
    bioremediation process, 218
    functional metagenomic approach, 215, 216f
    genomes and reconstruction of phylogeny, 216
    human health, 217–218
    industrial applications, 218
    methodology, 214–215
    microbial communities, 216
    microbial ecology, 218
    microbiomes from humans, 217
    structural metagenomic approach, 215
Methylation, 6
Microbial communities, 216
Microbial ecology, 218
Microbial genome sequencing, 59–60, 60t
Model assessment, 168–169
Modeling of loops, 168
MUSTER, 169
Mycoplasma genitalium, 59–60, 60t

N
Neurological diseases, 212
Next-generation sequencing methods, 51–54
    Illumina (Solexa) sequencing, 51
    ion torrent, 52

Roche 454 sequencing, 52
sequencing methods, 53—54
techniques, 51—53
Nuclear magnetic resonance (NMR) spectroscopy,
    149—154, 198—199
    applications of, 154
    basic principle, 150—151
    chemical shift, 152—153
    π-electron functions, 153
    H_2O, molecule, 152
    magnetic field, 151
    metabolite structures, 154
    proton, 151—152
    signal strength, 153
    significance of, 154
Nutrigenomics, 203
    beginning of, 92
    celiac disease, 94—95
    food allergies, 96
    functional foods, 97—104
        Amaranth, 100
        Buckwheat, 99
        crop, 102
        food quality, 102
        lifestyle diseases, 102
        lifestyle diseases, plant-based foods to,
            100—101
        Quinoa, 99
        regulating hypertension, 101
        technical intervention, 103—104, 103f
    gastroesophageal reflux disease (GERD), 95
    human nutrition and diseases, 94
    irritable bowel syndrome, 95
    lifestyle-associated diseases, 96—97
    oral disease, 96
    research tools, 92—93
    single nucleotide polymorphism, 93—94
Nutrition, 213

O

Oral disease, 96
Organelle genome editing, 18—20
    mitome-mitochondrial DNA, 19—20
    plastome-chloroplast DNA, 19

P

Personalized medicine, 203
Phage display, 180—181
Pharmaceuticals analysis, 146
Phyre, 169
Physical mapping, 40—44
    Fluorescence in situ hybridization, 41—43

restriction mapping, 40—41
Physical maps, 30
Polyacrylamide gel electrophoresis (PAGE)
    buffers, 114
    gel, 114—115
    gel and visualization of proteins, 116
    gel interpretation, 116
    gel running, 116
    polymerization of, 114—115, 115t
    principle of, 111—113, 112t
    protein samples, preparation of, 115—116
    requirements, 113—114
    sodium dodecyl sulfate (SDS), 111—116
    workflow, 113—114
Polymerization, 114—115, 115t
Posttranslational modifications, proteins on, 134
PRODORIC2 database, 81
Protein chips, 182—184, 184f
Protein Databases, 167
Protein—DNA interactions, 186
Protein—drug interactions, 186
Protein—lipid interactions, 185—186
Protein microarrays, 184—186
    analytical microarrays, 184—185
    functional protein microarrays, 185—186
Protein—protein interactions, 177, 178f, 185—186
Protein samples, preparation of, 115—116
Protein secondary and tertiary structure predic-
    tions, 165—166
Protein structure prediction, 165
Protein threading, 169—170
    software for, 169—170
Proteome expression, changing dynamics in, 133
Proteomes analysis-I
    2D gel electrophoresis-combining SDS-PAGE,
        118—120
        basic principle, 118
        fitting and loading first-dimension gel (IEF),
            120
        gel for one-dimension IEF process, 118—119
        preparing and pouring of, 119—120
    isobaric tags for relative and absolute quantifica-
        tion (iTRAQ), 128—130
        advantages, 130
        applications, 130
        basic principle, 129—130
        disadvantages of, 130
    isoelectric focusing gels (IEF), 117—118
        preparation, 117—118
    isotope-coded affinity tags, 125—127
        applications of, 127
        basic principle, 125—126

Proteomes analysis-I (*Continued*)
    workflow, 125−126
  isotope-coded protein labeling, 127−128
    applications, 128
  MALDI-TOF mass spectrometry, 120−125
    bioimaging, 124
    biomarkers, 124
    higher resolution reflectron, 124
    laser types, 122
    liner TOF analyzer, 122
    matrix compounds, 122
    microorganisms, 124
    polymers, 125
    principle and setup of, 120
    proteomes, 124
    reflectron TOF analyzer, 122
    sample concentration for, 121
    sampling plates, 121−122
  polyacrylamide gel electrophoresis (PAGE)
    buffers, 114
    gel, 114−115
    gel and visualization of proteins, 116
    gel interpretation, 116
    gel running, 116
    polymerization of, 114−115, 115t
    principle of, 111−113, 112t
    protein samples, preparation of, 115−116
    requirements, 113−114
    sodium dodecyl sulfate (SDS), 111−116
    workflow, 113−114
  quality control, 125
  stable isotopic labeling of amino acids in cell
     culture (SILAC), 131−134
    cancer cell proteomes, 134
    posttranslational modifications, proteins on,
     134
    principle of, 132
    proteome expression, changing dynamics in,
     133
    secretomes, 133−134
    technical workflow, 132−133, 134f
  tandem mass tag, 131
Proteomes analysis−II
  applications, 146
  bacterial cells, 146
  beverages analysis, 146
  circular dichroism (CD), 159−161
    applications, 161
    basic workflow, 160
  DNA and RNA analysis, 145
  fluorescence detection workflow, 147−148
    filter fluorometer, 147−148

    major applications of, 148−149
  fluorescence spectroscopy, 146−149
    principle, 146−147
  nuclear magnetic resonance (NMR) spectroscopy,
    149−154
    applications of, 154
    basic principle, 150−151
    chemical shift, 152−153
    π-electron functions, 153
    H_2O, molecule, 152
    magnetic field, 151
    metabolite structures, 154
    proton, 151−152
    signal strength, 153
    significance of, 154
  pharmaceuticals analysis, 146
  ultraviolet and visible light spectroscopy,
    139−146
    basic principle, 140−141, 142f
    basic workflow, 143, 143f
    monochromator, 143−144, 144f
    samples analysis, basics for, 144
    strengths/limitations, 145
    UV/Vis spectrum absorption, 141−142
  X-ray diffraction, 154−159
    applications of, 158−159
    crystal structure determination, 157
    mechanism, 156−157
    proteins, crystallization of, 155−156
    retrieving and processing of data, 158
    rotating crystal method, 157
    X-ray crystallography, 156
Proteomes analysis-III
  Ab initio protein structure predictions, 170−172
  generation of models, 167
    fragments assembly, 167
    segments, matching of, 167
  homology/template-based modeling, 166−169
  model assessment, 168−169
  modeling of loops, 168
  protein secondary and tertiary structure predic-
    tions, 165−166
  protein structure prediction, 165
  protein threading, 169−170
    software for, 169−170
Proteomes analysis-IV
  application, 181−182, 183f
  future prospectives, 187
  phage display, 180−181
  protein chips, 182−184, 184f
  protein−DNA interactions, 186
  protein−drug interactions, 186

protein—lipid interactions, 185—186
protein microarrays, 184—186
   analytical microarrays, 184—185
   functional protein microarrays, 185—186
protein—protein, 185—186
protein—protein interactions, 177, 178f
yeast 2-hybrid system (Y2H) approach, 177—179,
   179f
   advantages, 180
   limitations, 180
   principle, 179
Proteomics, 21f
   central elements of, 23—24
   primary structure, prolog to, 22—23
   protein structure, 21—25
   proteome profiling, 25—26
   quaternary structure, 25
   secondary structure, 23—24
      α Helix, 23—24
      β Helix, 24
      smaller secondary—βturns, 24
   structural hierarchies, 21—25
   tertiary structure, 24—25

Q

Quinoa, 99

R

Randomly amplified polymorphic DNA, 33—35,
   34f
   advantages, 34
   applications, 34
   limitations, 35
RAPTOR, 169
RedChIP, 79—80
Regulatory elements, 12—18
   CpG islands, 13
   highly repetitive DNA sequences, 15
   intergenic/extragenic DNA, 14
   introns, 14
   long interspersed elements, 18
   LTR transposons, 17
   moderately repetitive DNA sequences,
      16—18
   non-LTR transposons, 17—18
   nonrepetitive and repetitive DNA sequences,
      14—15
   other noncoding sequences, 13
   others, 13
   promoters, 12—13
   pseudogenes, 14
   5' untranslated region (UTR), 14

Restriction fragment length polymorphism,
   35—36
   application, 36
RNA polymerase II, 81—82
Roche 454 sequencing, 52

S

Sangers sequencing, 48—49
Secretomes, 133—134
Segment-matching approach, 167
Sequence-tagged site mapping, 44
Sequencing methods, 53—54
Simple-sequence repeats (SSRs), 31—32, 32t
Single nucleotide polymorphism (SNP), 37—40,
   39f, 93—94
Small ubiquitin-like modifier (SUMO), 7
Sodium dodecyl sulfate (SDS), 111—116
SPARKS X, 169
Stable isotopic labeling of amino acids in cell
      culture (SILAC), 131—134
   cancer cell proteomes, 134
   posttranslational modifications, proteins on, 134
   principle of, 132
   proteome expression, changing dynamics in, 133
   secretomes, 133—134
   technical workflow, 132—133, 134f
Structural genomics, 3—4, 4f
Structural metagenomic approach, 215

T

Tandem mass tag, 131
Target protein—DNA complex, immunoprecipita-
   tion of, 76
Technical workflow, 132—133, 134f
Transcription factors, 81—82
Transcriptomics, 192—196
   technological intervention, 192—196
      massively parallel signature sequencing
         (MPSS), 196
      microarrays, 192—194
      RNA-Seq, 194—195
      transcriptome-wide association studies
         (TWAS), 196

U

Ubiquitination, 6
Ultraviolet and visible light spectroscopy,
      139—146
   basic principle, 140—141, 142f
   basic workflow, 143, 143f
   monochromator, 143—144, 144f
   samples analysis, basics for, 144

Ultraviolet and visible light spectroscopy
(*Continued*)
strengths/limitations, 145
UV/Vis spectrum absorption, 141–142

W

Whole-genome shotgun sequencing, 50

X

X-ray diffraction, 154–159
applications of, 158–159
crystal structure determination, 157

mechanism, 156–157
proteins, crystallization of, 155–156
retrieving and processing of data, 158
rotating crystal method, 157
X-ray crystallography, 156

Y

Yeast 2-hybrid system (Y2H) approach, 177–179,
179f
advantages, 180
limitations, 180
principle, 179